Energy 2000

Energy 2000

A Global Strategy for Sustainable Development

A Report for the World Commission on Environment and Development

Zed Books Ltd
London and New Jersey

Energy 2000: A Global Strategy for Sustainable Development was first published on behalf of the World Commission on Environment and Development, Palais Wilson, 52 rue des Paquis, CH-1201, Geneva, Switzerland, by Zed Books Ltd., 57 Caledonian Road, London N1 9BU, UK, and 171 First Avenue, Atlantic Highlands, New Jersey 07716, USA, in 1987.

Cover designed by Adrian Yeeles/Artworkers.
Diagrams drawn by Henry Iles.
Typeset by Grassroots Typeset, London.
Printed and bound in the United Kingdom by
Cox & Wyman Ltd., Reading.

British Library Cataloguing in Publication Data

Energy 2000 : a global strategy for sustainable development: a report for the World Commission on Environment and Development.
1. Energy conservation
I. World Commission on Environment and Development
333.79'16   TJ163.3

ISBN 0-86232-710-5
ISBN 0-86232-711-3 Pbk

Contents

Tables and Figures

Foreword

In 1983 the United Nations General Assembly called for the establishment of a special Commission to look at the critical issues of environment and development from the perspective of the year 2,000 and beyond, and to propose better ways and means for the world community to address them.

The World Commission on Environment and Development first met in October 1984, and in 1985 it established a special Panel on Energy to analyse the main energy-related environment and development issues, and to recommend the types of measures needed to meet the world's future needs for energy on a basis that promise to be ecologically and economically sustainable through the turn of the century.

In 1985, the Chairman of the World Commission on Environment and Development, Prime Minister Gro Harlem Brundtland of Norway, asked me to chair the Panel, a task that I accepted as both a major responsibility and a great opportunity. I was joined by an excellent group of colleagues who together shaped, and on whose behalf I presented, the following report to the Commission at its Fifth Meeting in Ottawa on 28 May 1986.

The patterns of energy consumption and supply, and their implications for development and the environment through the year 2000, are being determined today. Unfortunately, today's energy choices by governments, industries and individuals, and the narrow terms of reference within which those choices are made, results in overall energy patterns which, in the aggregate, are not sustainable; neither ecologically, nor economically. Some involving the overuse of resources, such as biomass consumption in developing countries, transform renewable into nonrenewable resources; others involving the inefficient and unnecessarily dirty use of fossil fuels, cause avoidable air pollution in our cities and acidification of the environment, which is now restricting the development potential of several industrialized countries, and is beginning to appear in developing countries. Other patterns threaten massive changes in global climatic conditions, with enormous consequences for development and security.

As this report demonstrates, it is possible to develop energy strategies that avoid or minimize these syndromes and that enhance rather than diminish the potential for economic development. Such strategies would aim to secure a major shift from present to more productive patterns of energy consumption and improved forms of production, including a much greater emphasis on benign sources. They would include strategies to negotiate predictable world oil prices, at levels high enough to encourage efficiency measures; to reduce the use of fossil fuels wherever possible, and use them in environmentally favourable conversion systems; stabilize the woodfuel resource base in developing counries; increase the role of renewables substantially; and make nuclear energy an acceptable source of power.

The changes required to implement such strategies, however, face enormous obstacles—economic, institutional and political. The report considers these and concludes that, however difficult, government, industries and individual energy users must make a total commitment to a transition to more sustainable energy patterns. If we return to energy prices that favour high levels of energy consumption per unit of growth, and to policies that induce massive investments in traditional sources of supply, which are so wasteful of resources, pollute the environment and inhibit economic development, the world will lurch from avoidable crisis to avoidable crisis, both developmental and environmental.

Although energy strategies favourable to both environment and development will require major shifts in national policies, no single country has the resources to make these shifts on its own. At a time when multilateral co-operation is being neglected everywhere, concerted international action has become imperative. Measures to strengthen regional and international co-operation are essential, and it is my hope that those put forward here will stimulate discussion and enable the Commission to recommend a number of concrete measures when it reports to the General Assembly in 1987.

I would like to express my appreciation to my colleagues on the Panel for their generous efforts and wise counsel and the members of the Secretariat who worked so diligently to support the Panel.

Enrique Iglesias
Chairman
Advisory Panel on Energy

1. World Energy Prices, Environment and Development

When this Panel was established by the World Commission on Environment and Development in March 1985, world oil prices were at $28-30 per barrel and economic development everywhere was being planned on the assumption of a steady increase in the price of energy. As we completed our report in May 1986, world oil prices were around $10-12 per barrel and old prophets were announcing a new era of cheap energy.

With the breakdown of effective stabilizing factors, oil prices have become impossible to predict. Looking to the next decade and the year 2000 and beyond, it seems clear that, in the face of increasing economic activity and of constantly rising demand, the *supply* of this non-renewable resource will emerge as the decisive factor in determining prices. It is imperative, therefore, that a period of cheap hydrocarbon energy should not be taken as a permanent phenomenon. Short-term gains are always tempting to those in a position to grab them, whether producers or consumers. Such gains may temporarily set back, but they in no way alter the fundamental character of the transition now under way to a broader mixture of energy sources, that includes a steadily increasing proportion of renewables.

The world can stumble through this transition, passing from shock to shock at great cost to sustainable development and its environmental bases. Or it can manage the transition, using more sustainable mixtures of energy supply and patterns of energy consumption. This, however, will require a transformation in energy policies and institutions.

An effective arrangement to stabilize wild fluctuations in world oil price, at a reasonable level in real terms is essential to this transformation. In the absence of such an arrangement,

1

oil prices are bound to fluctuate over a wide range. In 1979, the time was ripe to conclude such an arrangement but the opportunity was lost because of myopia among the exporting and confusion among the importing countries. In 1986 the time is again ripe, but there is a very real danger that the opportunity will again be lost, this time because of myopia among the importing and confusion among the exporting countries.

The economic, social and environmental costs of losing this opportunity are much clearer today than they were in 1979.

In industrialized countries during the past decade economic growth has finally been decoupled from parallel growth in energy use. The incremental energy content of growth (that is, the new energy required for every new increment in growth) has fallen in many countries, in some from 1.2 to 0.5 units, resulting in substantial gains in overall economic efficiency and competitiveness and substantial reductions in the costs of environmental damage. The momentum that produced energy-efficiency gains of up to two per cent a year, because of high oil prices, is now threatened in many sectors—in transportation, industry, agriculture and others—and could quickly be lost.

Developments that made sense with oil at $25 per barrel make no sense at all with oil at around $10 per barrel. Many producers, including those involved in new ventures in renewable sources of energy and conservation, have been ruined, and massive investments in the search for new oil, and in the development of the renewable energy sources that will be needed through the transition, have been placed in temporary and perhaps terminal jeopardy.

Developments that need to be sustained to the year 2000 and beyond, including most major developments in transportation, agriculture, industry and other sectors, need to be based on a realistic view of the energy transition. Planning the future on the basis of cheap energy will recoil with vengeance on both development and equity when prices rebound, as happened in the 1970s.

The material and commodity content of growth has also fallen during the past decade because of the development of substitutes. A temporary period of cheap oil will enhance the position of oil-based substitutes, in textiles and rubber for example, and further reinforce this process. In the future, therefore, growth in developing countries will not benefit to the same degree from growth in industrialized countries.

Failure to re-establish world oil prices (and hence consumer prices of energy) at a level that sustains annual gains in energy efficiency and a steady shift to renewables will threaten many nations not only with reduced development potential, but also with reduced security. A period of unreasonably low prices will re-establish patterns of high energy supply and consumption, with a higher dependence on fossil fuels. These patterns will reinforce the processes of trans-boundary air pollution and environmental acidification now under way in Europe and North America and will hasten their spread to other regions in Asia, Africa, and Latin America. They will also accelerate the processes of climatic change now under way. The scientists concerned recently warned that within the next 40 to 60 years climatic change could provoke not only a major shift in climatic zones but also rises in sea level of 145 centimetres, thus adding a steady shift in the boundaries of coastal states to the agenda of international tensions.

Effective attempts to stabilize wild fluctuations in world oil prices will no doubt take some years to negotiate. In the meantime, nations may choose to allow consumer prices to fall to levels dictated by the market or they may deploy various measures to sustain prices at higher levels.

At the moment, the former course is generally being pursued. With oil prices falling to around $10 a barrel, and the subsequent prospect of oil at pre-1973 prices in real terms, confident talk of the market soon finding its own *natural floor* has been silenced. Extreme price uncertainty is leading to unbridled market speculation in oil futures, which is imposing severe economic and social hardship on the oil-exporting nations. In the longer term, with unrestricted demand gradually overtaking supply, the stage will be set for yet another energy shock, and a repeat of the economic, social and environmental experience of the 1970s.

On the other hand, consumer nations, especially major industrialized countries, could take steps to maintain prices and through their budgets retrieve a major portion of the wealth created by the sudden collapse of world oil prices. This would provide a major source of revenue, enable governments to

3

reduce their heavy debt burden and, return to sounder fiscal management. It would ensure that the momentum producing more sustainable patterns of energy consumption and supply is not lost and that annual gains in energy efficiency continue. It would reduce the rising cost of damage from urban air pollution and the acidification of the environment, and it would buy the time needed to develop and apply strategies to reduce and/or adapt to climate change. Moreover, in thus aggregating a portion of the windfall of more than $100 billion that would flow to their consumers, industrialized countries would be in a position to increase substantially their contributions to bilateral and multilateral agencies in supporting developing countries to establish the institutions and effect the policies and environmental regeneration needed to manage the energy transition.

We recognize, however, that action to impose higher consumer prices would generally be extremely difficult owing to domestic political and international economic constraints. Perhaps the only genuine solution lies in the steadily growing realization that oil is such a crucial raw material to future strategies for environment and development that it is too important to be treated as just another commodity and left to the vagaries of a volatile world market. Instead, both producers and consumers should co-operate in developing policies towards building some limited form of global convention for the more orderly production and marketing of oil.

2. The Strategic Choices

The patterns of energy supply and consumption of the year 2000 and beyond are being determined today. The same is true, of course, of the patterns of population growth, industrialization, urbanization, agriculture and transport. But energy patterns are unique. They can widen the opportunities for future economic and social development or they can constrain them, in some cases eliminating them altogether. They can provide the basis for a high and sustainable level of security and comfort; or they can destroy it, reinforcing insecurity, widespread poverty and human misery. Ultimately, they can determine the capacity of this planet to sustain the life support systems on which all other development depends.

There is no doubt that today certain energy patterns threaten the sustainable development of the countries and regions in which they are dominant. This is clearly true of the fuelwood crisis: according to an FAO study[1], in 1980, 1,300 million people lived in wood deficit areas (defined as areas where people can still satisfy minimum needs, but only through unsustainable overcutting) and over 110 million in acute scarcity areas (defined as areas where even with overcutting people cannot satisfy minimum needs). The same study projects that by the year 2000 about 3,000 million people may live in wood deficit and acute scarcity areas, thus causing increased deforestation, erosion, desertification and the diversion of plant and animal wastes from soil replenishment (See Table 1).

Sustainable development is also threatened by the acidification of the environment in Europe and North America, a phenomenon now spreading to parts of Asia, Africa and Latin America. The silent accumulation of acid has led to the wide-

5

GLOBAL DISTRIBUTION OF PRIMARY ENERGY USE

1985

WORLD

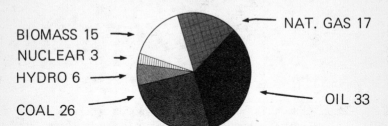

BIOMASS 15 →
NUCLEAR 3 →
HYDRO 6 →
COAL 26 →

← NAT. GAS 17

← OIL 33

INDUSTRIALIZED COUNTRIES

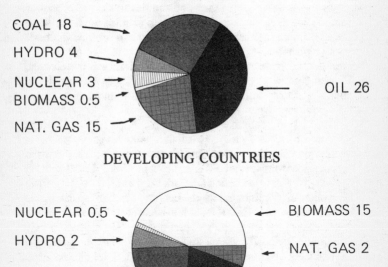

COAL 18 →
HYDRO 4 →
NUCLEAR 3 →
BIOMASS 0.5 →
NAT. GAS 15 →

← OIL 26

DEVELOPING COUNTRIES

NUCLEAR 0.5 ↘
HYDRO 2 →
COAL 8 →

← BIOMASS 15

← NAT. GAS 2

← OIL 7

*Source: BP - 1985 BP Statistical Review of World Energy,
British Petroleum, UK, June 1985.*

6

Table 1
Population Experiencing a Fuelwood Deficit
1980 and 2000 (in millions)
(Source WRI—1986)

Region	1980 Acute Scarcity		Deficit		2000 Acute Scarcity or Deficit	
	Total Population	Rural Population	Total Population	Rural Population	Total Population	Rural Population
Africa	55	49	146	131	535	464
Near East and North Africa			104	69	268	158
Asia and Pacific	31	29	832	710	1,671	1,434
Latin America	26	18	201	143	512	342
Total	112	96	1,283	1,052	2,986	2,398

Note: Total population and rural population (total population less that of towns with more than 100,000 inhabitants) in zones whose fuelwood situation has been classified.

Source: Adapted from Food and Agriculture Organization, 1983. Reference 67.

spread sterilization of lakes and, more recently, of soils. The latter may be the principal cause of the accelerating death of Europe's forests, the front line in the security system of any community, nation or region. Without them, a nation's economy is naked, its soil exposed to rapid run off and erosion, its valleys to flooding and its valley homes and agriculture to destruction. In October 1985, Switzerland was reeling from a decision to evacuate 70 people from a mountain village, perched below a forest that had been killed by acid rain. In 1986, they may need to evacuate over 100,000 people. Here, tomorrow has become today.

Certain patterns of energy consumption could undermine sustainable development on a global basis. Climatic change has now emerged as a *serious and plausible* threat to the economic and social development of nations and could place intolerable burdens on worldwide political stability within the next 40 to 60 years.

Fortunately, however, there are choices open to most nations and to the international community that could reduce, if not prevent, the threats that certain types of energy patterns pose to sustainable development. Some of these choices hinge on policies to reduce the future demand for energy without reducing the potential for economic growth. Others hinge on policies to manage better the burning of fossil and other fuels, internalizing the enormous costs they now impose on their own community and on other nations and regions. Still others hinge on policies to induce a steady shift in types of energy from nonrenewable to renewable sources. All require changes in existing policies and new and strengthened institutions at the national and international level.

A High or Low Energy Future

The major strategic choice before governments and the world community is illustrated in Figure 2, which includes some of the better known projections[2] of energy use up to the middle of the next century. It will be seen that by the year 2000 global energy consumption varies by a factor of five between the lowest and the highest projections, and by the year 2020 the spread

PROJECTIONS OF PRIMARY ENERGY CONSUMPTION

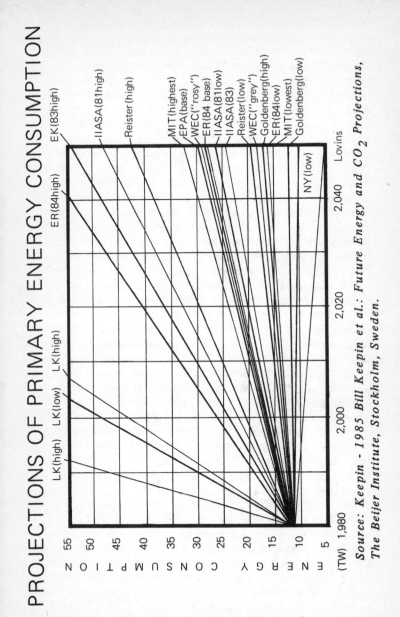

Source: Keepin - 1985 Bill Keepin et al.: Future Energy and CO_2 Projections, The Beijer Institute, Stockholm, Sweden.

9

between the projections is too large to portray here.

In examining Figure 2, it is vital to understand that *energy use projections are not predictions* of the future. They are projections of possibilities, useful only to analyse what can happen under different assumptions concerning all the factors that together determine the pattern of energy supply and consumption. The differences in Figure 2 thus reflect not so much differences in methodology as different assumptions which will influence energy patterns, especially gains in energy productivity.

Given the rich range of projections now available in various studies, we have not attempted to add yet another of our own. Instead, and for the sole purpose of illustrating the energy, economic and environmental implications of possible different energy futures, two of these projections have been selected, to represent respectively a credible *upper limit* and *lower limit*.

Our purpose in employing these projections, therefore, is not to evaluate and defend the absolute numbers involved. Rather is it to show that two radically different energy futures, reflecting different policies and institutions, are possible. The projections also provide parameters within which to discuss the consequences of two broad directions for development and for sustaining the environmental bases of development.

The high scenario indicates the direction in which energy consumption and supply patterns are heading if existing policies and institutions remain essentially unchanged. It is a trend scenario, included in a study published in 1981 by the International Institute for Applied Systems Analysis[3]. It assumed world economic growth at an average annual rate of 2.1 per cent. With no gains in energy efficiency, indeed with energy efficiency on a global basis falling at a rate of one per cent per year, it projects a tripling of global energy consumption over 1980 levels by 2020.

The most recent examination of the technical feasibility of reaching a low energy future is that contained in a 1985 study by an international group of energy analysts[4]. It is not strictly an energy projection, but rather a feasibility study, demonstrating what could be achieved if all future development incorporated the most energy-efficient technologies and processes now available and in use in housing, industry, transportation and other sectors. Assuming this were possible, the study projects a 50 per cent drop in per capita energy consumption in industrialized and only a 30 per cent increase in developing countries.

These and other assumptions produce a mere ten per cent increase in global energy consumption by 2020, a striking figure when we recognize that it is compatible with economic growth rates similar to those employed to project the high scenario. In fact, no GDP growth rate was explicitly assumed in the projection. According to the authors, however, it is consistent with an average annual growth rate of 1.7 per cent in industrialized countries and up to 6 per cent in developing countries. It is also consistent with an average annual growth rate globally of up to about 3 per cent per year.

Taken together, these assumptions imply that average annual gains in energy efficiency of 3.3 per cent and 2.7 per cent can be reached and sustained in industrialized and developing countries respectively. While this may be technically feasible, it presumes very high rates of penetration by energy efficient technologies and processes. Without in any way diminishing the value of the study, we feel that somewhat less optimistic assumptions would be more reasonable, given past performance and well-known economic, social and institutional constraints. Figure 3 shows data about the past performance of the OECD region. On the average, the countries concerned have achieved annual energy efficiency increases of 1.7 per cent per annum between 1973 and 1983. For them to achieve rates as high as 3.3 per cent would require the rate of penetration of new technology to increase substantially, an unlikely event when the world is facing a period of declining real prices for oil.

In the case of developing countries, a lower energy scenario requires not just a rise in the rate of penetration of new technologies, but also a reversal of current trends. According to a World Bank report[5], in the 1960s and the 1970s the energy content of GNP in developing countries actually increased by well over two per cent a year. A reduction of 2.7 per cent in the future would constitute a departure of 4.7 per cent from current trends.

In view of this, we feel that it would be more reasonable to expect the energy future to unfold somewhere between these

12

TOTAL ENERGY REQUIRED PER UNIT OF GROSS DOMESTIC PRODUCT

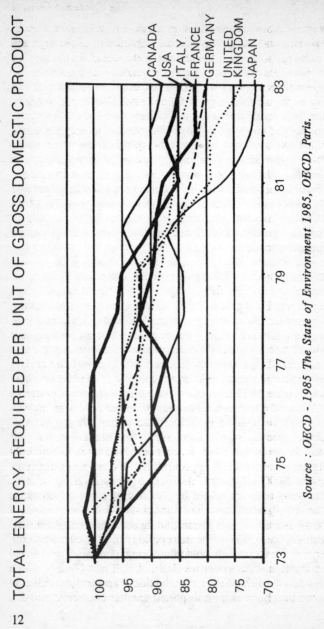

Source : OECD - 1985 The State of Environment 1985, OECD, Paris.

two projections. Nonetheless, given the interrelated economic, environmental and developmental implications of the high and low energy scenarios, the nations of the world should aim for the lowest possible energy future compatible with their economic and social priorities. We recognize, however, that while this may be the ultimate aim, if the lowest option is not possible, then, reluctantly, a higher would have to be accepted.

In view of this, we felt that if the low energy scenario could actually be achieved, it would be sufficient to target future policies on increasing the energy content of development and controlling the emission of pollutants. But if, as seems likely, substantial increases in energy consumption will take place, then all countries, and in particular developing countries, will have to explore urgently all methods of increasing the supply of energy in environmentally benign ways, particularly by using renewable sources.

Energy supply implications
The energy supply implications of the high scenario are staggering. By the year 2020, oil and natural gas would have to be produced at almost twice the 1980 rate, while coal production would need to increase by a factor of 1.8. This increase in fossil-fuel use implies a capacity equivalent to bringing a new Alaska Pipeline (two million barrels of oil equivalent per day) into production every one to two months! (It should be noted that even the 1980 level of oil consumption carried through to 2020 would require the discovery of 20 per cent more proven reserves than existed in 1985.[6]) More than six terawatts of nuclear capacity would also have to be installed by the year 2020, an increase of 3,000 per cent over 1983 levels! That would require commissioning approximately 150 large nuclear reactors of 1,000 megawatts electrical capacity each per year.

Existing reserves would be able to cover the requirements for natural gas and coal, but it is questionable if the necessary infrastructure for mining, transporting and converting these fuels could be developed in this short period in all the countries concerned. For example, about 90 per cent of all reserves of coal are found in three countries (USA, USSR and China), which would have to double their production. Furthermore, this coal would then have to be transported to other countries, and this

13

would require an enormous increase in rail, pipeline and terminal handling capacity.

With the low scenario, the implications on energy supply are still considerable, but much more manageable. Oil and coal would be used at rates approximately 20 per cent less than in 1980, and only nuclear, hydro and natural gas would increase above 1980 levels by factors of 2.4, 1.4 and 0.8 respectively. Since new additions to the supply of a particular energy source are proportionately much more expensive than existing sources, the potential gain by not having to open up new supplies could be very large indeed. The main implications of the low scenario, however, are for the management of energy demanded.

Economic and developmental implications

The economic implications of the high scenario are also substantial. The investment requirements are so enormous that even the industrialized countries would find them difficult to support. In developing countries, the high level of investment required would siphon off capital required for other sectors, which would be starved. In fact, many now believe that these requirements would be a drag on development and make it impossible to improve living standards in developing countries.

According to World Bank estimates, $130 billion would have to be invested every year in energy projects in order to raise per capita levels of commercial energy consumption in developing countries from 0.54 to 0.78 kW between 1980-1995 (which is necessary to reach the high projection by 2020). Moreover, half of this, that is $65 billion (compared to the $3.5 billion currently being loaned in the energy sector) would be in foreign exchange, the equivalent of about four per cent of aggregate GNP of these countries.[7] This level of investment, involving such a huge charge on foreign exchange earnings, would push most energy importing developing countries further into the already serious debt crisis.

The economic implications of the low scenario, on the other hand, could be beneficial. While achieving more or less the same level of economic growth as the high scenario, it does this with a much lower level of energy consumption. Not only are investment requirements and corresponding debt burdens lower, but also a greater reliance on renewable forms of energy would have a much higher employment impact, especially in rural areas. Lower levels of conventional energy production would also reduce the investment required to prevent or control environmental damage. Greater reliance on renewables, if properly managed, could lead to significant opportunities for environmental regeneration and provide an economic base for reforestation, wasteland reclamation and other measures.

Although it is not possible to quote a figure for the overall investment requirements of the low scenario, as was done for the high scenario, it can be shown that, under a wide range of circumstances, the extra capital requirements for more efficient technologies will be more than offset by capital savings for lowered energy needs[8]. For example, in the case of Brazil, it has been shown that for a discounted, total investment of four billion dollars in more efficient technologies (e.g. better refrigerators, lighting, motors, etc.) it would be feasible to defer construction of 21 gigawatts of new electrical supply capacity; this corresponds to a discounted capital savings for new supplies of $19 billion in the period 1986 to 2000.[9]

Similarly, for a new advanced steel-making process in Sweden the overall investment requirements are in fact lower than for conventional ones and accompanied by savings of 50 per cent or more in energy consumption.[10] Again, the difference saved could be invested in some other sector of the economy.

Environmental implications
The high scenario, with its heavy reliance on fossil fuels and its huge requirements for investment in other conventional sources of energy, has equally staggering environmental implications.

In terms of fossil fuel use, for example, the high scenario would result in substantial increases in consumption of coal, oil and natural gas. The increase in the first of these, coal, is most worrisome, since the environmental and health effects of coal are greater than those of oil. Overall, there would be a more than doubling of the production of carbon dioxide, which would bring the world much closer to possible major climatic changes. Increased fossil fuel combustion in power stations as well as in automobiles would also aggravate the acid rain problem in industrialized countries and developing countries alike.

In developing countries, the indirect environmental impacts of such a high energy use scenario could be disastrous. On the one hand, local energy supplies would have to be substantially increased. Carelessly installed energy development projects, such as the construction of large dams, have greatly contributed to land degradation and consequent impact on the environment.

On the other hand, the high energy costs implied by such a future will require energy importing developing countries to step up commodity production for export (as opposed to production for local consumption) to be able to pay for the higher debts incurred. More land will have to be devoted to non-food, export-oriented agricultural production; this means either the opening up of more forests (i.e. deforestation) and/or the marginalization of even more subsistence farmers onto marginal, lower quality lands. In both cases, the environment will suffer in the process. Deforested land is often subject to misuse which results in eroded topsoils. Marginalized farmers are obliged to overuse the low quality soils, with the same results: soil erosion, silting up of reservoirs downstream, and eventually floods. In all cases, the impacts have tremendous developmental costs, for which society as a whole has to pay.

With the *low* scenario, a large number of environmental effects to be avoided or abated will remain. There will, however, be no increases above the 1980 levels. In fact, although fossil-fuel use will remain essentially constant, there will be a small drop in the rate of carbon dioxide production owing to changes in the fossil-fuel mixture (i.e. more natural gas and less coal and oil). This will slow down the rate of climatic change and give more time for the world community to deal with the problem.

Reducing Fossil Fuels

Although all energy systems have undesirable environmental effects, high levels of fossil-fuel consumption are a special concern, particularly in industrialized countries and in the industrialized and urban regions of developing countries. They pose three interrelated threats to sustainable development: air pollution,[12] acidification of the environment[13] and climatic change.[14] Some industrialized countries may possess the

Table 2
Energy Imports as a Percentage of Merchandise Exports in Developing Countries (Source: Miller-1986)

Country	1983	Country	1983
Central & S. America		Africa	
Argentina	9%	Algeria	2%
Brazil	56	Burkina Faso	50
Chile	24	Cameroon	4
Colombia	21	Egypt	12
Costa Rica	22	Ivory Coast	16
Dominican Republic	71	Madagascar	32
El Salvador	57	Morocco	57
Guatemala	68	Niger	17
Honduras	28	Senegal	58
Nicaragua	46	Sudan	57
Panama	82	Togo	18
Paraguay	1		
Peru	2	Asia	
Trinidad & Tobago	4	Bangladesh	20%
Uruguay	28	Hong Kong	7
Venezuela	1	Indonesia	20
		Korea, Rep. of	28
Europe		Malaysia	16
Greece	59%	Pakistan	49
Portugal	48	Philippines	44
Turkey	66	Singapore	40
Yugoslavia	33	Sri Lanka	40
		Thailand	39
		Middle East	
		Jordan	101%

Source: World Bank *World Development Report 1985*

economic and social resilience and institutional capacity needed to cope with these threats, but most developing countries do not. In developing countries the principal threat is the foreign exchange costs of imported fuels (See Table 2) and of the pollution control equipment. Moreover, in many of them, the high energy scenario would impose even greater direct damage to their resource base, accelerating deforestation, erosion, siltation and flooding and reinforcing the fuelwood crises. We will come back to this after a brief look at the fossil-fuel implications.

Air pollution

During the past three decades of rapid growth, urban air pollution has increased dramatically, more or less in pace with fossil-fuel consumption for space heating (and cooling), automobile transport, industrial activities and power generation. Beginning in the late 1960s, a growing awareness of the effects of polluted air on human health, property and the environment created a demand for action. Some industrialized countries responded, enacting control measures of various kinds. Most imposed standards that resulted in the development of curative measures, including add-on technologies. Some imposed liability and required compensation for damage, especially damage to human health. While expensive, this react-and-cure approach led, in time, to reduced emissions of some of the principal pollutants and cleaner air over some cities. London, Tokyo, Montreal and New York are a few of the better known success stories. In time, it also induced innovation in the development of environmentally efficient technologies for household cooking and space heating, for industrial processes and for power generating.

Many industrialized countries, however, and virtually all developing countries failed to share in this experience. Instead, they witnessed a steady deterioration in the quality of their air with all its attendant effects. Air pollution has reached dramatic levels in most major Third World cities, far exceeding the worst cases of the 1950s in western industrialized countries. Sao Paulo, Rio de Janeiro, Buenos Aires, Lagos, New Delhi, Bangkok, Seoul and many other cities feature in recent studies. Mexico City, is known as one of the worst air pollution cases in the world.

The fossil-fuel emissions of principal concern include sulphur dioxide, nitrogen oxide, carbon monoxide, various hydrocarbons, fly ash and suspended particulates. These attack human health, damaging body tissue and the nervous system, bringing increased respiratory diseases and cancer, and causing higher morbidity and mortality in sensitive segments of the population. Transformed into acid, they burn and kill vegetation, corrode buildings and vehicles and eventually contribute to land and water pollution. Except for a few western industrialized countries, studies of the social and economic costs that these effects impose on the economy of communities and nations are non-existent or non-available. The few studies available, however, demonstrate that the effects are enormous, and in most of the world they are growing rapidly.

The sources of air pollution vary from city to city. So do the impacts, given each city's unique natural setting, altitude, climate and weather patterns. Factories, industrial plant and other fixed activities are a major source. The heavy oil burnt for household heating and thermal power generation is a leading cause of sulphur pollution. Trucks and automobiles contribute heavily to emissions of nitrogen oxide, carbon dioxide and hydro carbons. In many Third World cities, a thick film of smoke from burning raw coal, wood, and animal dung for cooking and heating provides a constant reminder of the toxic nature of the air being breathed.

Today, the fossil-fuel sources of air pollution are largely controllable in both industrialized and developing countries. Most are preventable, and at a cost to the community and nation that is usually far less than the costs of repairing damage that will otherwise be incurred, even without the eventual costs of retrofitting vehicles, homes and industry, when the effects of air pollution exceed the limits of community tolerance.

Under the high energy scenario, however, both prevention and control would be extremely difficult and expensive. Indeed, given projected economic trends on the one hand, and the state of awareness, legislation and institutional capacity in most developing countries on the other, it is doubtful whether they could keep up with future sources of pollution, let alone catch up with present sources and cure past damage. For similar reasons, this may well be beyond the capacity of many industrialized countries.

The most cost-effective means of prevention available, perhaps the only ones available to some industrialized and all developing countries given the high cost of pollute-and-cure processes, are those implicit in the low energy scenario. Moreover, these and other means would reinforce those needed to prevent and control another major fossil-fuel threat to sustainable development: acidification of the environment.

Acidification of the environment

The measures taken by industrialized countries in the 1970s to control urban and industrial air pollution (high stacks, for example) very often simply transferred the problem elsewhere, to other media or, more seriously in the short run, to the hinterland in their own and other countries. This was manifest in a rapid rise in transboundary air pollution in Europe and North America and in widespread acidification of the environment. In consequence, perception of air pollution has shifted from that of a local problem, involving one or more communities, to that of a regional problem involving entire continents.

During long-distance transport in the atmosphere, emissions of sulphur dioxide, nitrogen oxide and volatile hydrocarbons are transformed into sulphuric and nitric acids, ammonia salts and ozone. They fall to the ground, sometime thousands of miles from their point of emission, as dry particles or in rain, snow, frost, fog and dew.

Silently accumulating over the decades, the damage to the environment first became evident in Scandinavia in the 1960s and has since mounted at an accelerating pace. Several thousand lakes in Europe and North America have registered a steady increase in acidity levels to the point where they no longer support fish life. The same acids attack stonework and corrode metal structures causing damage estimated at billion of dollars annually. They enter drinking water supplies, liberate potentially toxic metals such as cadmium, lead, mercury, zinc, copper and aluminium, and pose risks to human health.

The evidence underlying the urgent need for action on the sources of acid rain is mounting with a rapidity that exceeds the capacity of scientists and governments to assess it.[14] Until now, the greatest damage has been reported over Central Europe, which is currently receiving more than one gram of sulphur on every square metre of ground each year (see Figure 4). There was little evidence of tree damage in Europe in 1970. In 1982 the Federal Republic of Germany reported visible damage to eight per cent of its trees; in 1983 this rose to 34 per cent and in 1985 to 50 per cent.[15] Sweden reported light to moderate damage in 30 per cent of its forests; and reports from other countries also become extremely disquieting. So far an estimated five to six per cent of all European forest land is affected.

Have European soils reached a danger point? The evidence is not yet complete, but many reports consistently show soils in parts of Europe as becoming acid throughout the tree rooting layers (c. 100cm deep). The acidity is frequently above this level, where aluminium enters into solution as a mobile element, and is toxic in very low concentrations to plant roots. Forest death could be caused by the simultaneous action of acid soils containing mobile aluminium and by direct needle damage in conifers from the interaction of the pollutants mentioned above.

If this is true, we may be witnessing in Europe an immense, regional acid-base chemical titration with potentially disastrous results being signalled by widespread tree damage and death. Trees may be in effect, a kind of *environmental litmus paper* which indicates a change to irreversible acidification whose remedial costs are beyond economic reach. Comparatively speaking, forest death on a regional scale would be socially and economically trivial compared to such consequences as erosion, siltation, flooding of farmlands and towns and local climatic change.

The effects of air pollution, such as forest death, acidification of lakes and damage to structures is evident. The sources are multiple, including ozone and sulphur dioxide, deposition of excess nitrogen-nitrate-ammonia, and deposition of acids and heavy metals. No single pollutant control strategy is likely to be effective in dealing with forest decline—it will take nothing less than a total integrated mixture of strategies and technologies, tailored for each region, to improve air quality significantly.

Evidence of acidification in the newly industrializing countries of Asia, Africa and Latin America is beginning to emerge. China, Korea and Japan seem particularly vulnerable, given

ACID DEPOSITION ON EUROPE

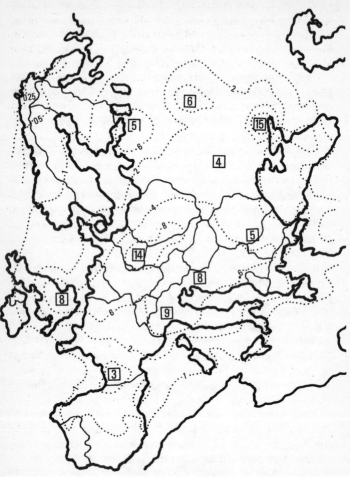

(Isolines of Average Annual Total Deposition of Sulphur, expressed as grams of sulphur per square metre of ground per year, based on the period October 1978 - September 1982. Maxima are shown as boxed numbers.
Source : adapted from EMEP/MSC - W Report 1/85).

industrialization trends in the former two, as do Venezuela, Colombia, Ecuador and Brazil. So little is known about the likely environmental loading of sulphur and nitrogen and about the capacity of tropical lakes and forest soils to neutralize acid that, at a minimum, a comprehensive programme of investigation should be formulated without delay.[16]

Developing countries cannot afford to repeat the mistakes made in Europe and North America. In the US, it has been estimated that reducing sulphur dioxide emissions by half from existing sources would cost three to four billion dollars a year, and increase electricity rates by two to three per cent. If nitrogen oxides were included, the costs might be as high as $6 billion dollars a year.[17] Estimates of the annual costs of securing a reduction in sulphur emissions in the European Community of 55 to 65 per cent between 1980 and 2000 range from $4.6 to $6.7 billion (1982) per year. Controls on stationary boilers to reduce nitrogen levels by only ten per cent per year by 2000 range between $0.1 and $0.4 billion (1982). While high in absolute terms, even in countries whose energy systems depend heavily on coal-based thermal power, these figures translate into a one-time increase of about six per cent in the price of electrical power to the consumer.

Estimates of the costs of damage in Europe are less reliable and, given the trends noted above, necessarily very conservative. Nonetheless, studies place costs due to material and fish losses alone at three billion dollars a year; while damage to crops, forests and health are estimated to exceed ten billion dollars per year. Again, the evidence is not yet complete. Recent Japanese laboratory studies indicate that air pollution and acid rain can reduce some wheat and rice crop production, perhaps by as much as 30 per cent. Given population trends and the needs for future food security, these are costs that no developing country would wish to contemplate.

Yet, in pursuing the high energy scenario, developing countries will inevitably incur substantial acidification with its attendant damage costs and threats to their sustainable development. Industrialized countries must either continue to accept these costs or adopt expensive retrofit measures. The dilemma of deciding who should bear the cost of curative measures, and when and how, has divided the nations of North America and Europe and driven acid rain to the top of their agendas for regional action.

The strategies implicit in the low energy scenario offer the most cost-effective means of reducing future levels of acidification in industrializing and developing countries alike. And they would buy time for the nations of the world to assess and prepare for the implications of global climatic change.

Climatic change

Upon combustion, fossil fuels emit the gas carbon dioxide, which accumulates in the atmosphere. The pre-industrial concentration was 280 parts of carbon dioxide per million parts of air by volume. This concentration rose to 340 in 1980 and is expected to reach 560 between the middle and the end of the next century. How soon this occurs will depend almost entirely on the energy paths followed in the meantime. In contrast to the emissions mentioned above, no technologies exist to control the emissions of carbon dioxide.

Other gases are also accumulating in the atmosphere, principally, chlorofluorocarbons (used as aerosol propellants in spray cans and in refrigerators as a coolant); methane (rising from wet, reducing soils, e.g. rice-paddies, or from the earth's surface, especially where oil or gas is exploited); nitrous oxide (derived from the breakdown of nitrogenous fertilizers); and ozone (generated by industry and internal combustion engines).

The question of climate warming caused by rising concentrations of these *greenhouse* gases in the atmosphere has been the object of intense assessment, nationally and internationally. The question is enormously complex given the interactive nature of the meteorological, oceanographic and ecological factors conditioning climate, ecosystem and sea level responses.

However, after reviewing the latest evidence in October 1985, scientists from 29 industrialized and developing countries concluded that climatic change must be considered a *plausible and serious possibility*.[18] They estimated that the concentration of greenhouse gases in the atmosphere could lead to a rise in global mean temperatures in the first half of the next century *greater than any in man's history*. A global warming of 1.5°C-4.5°C at the equator (greater at higher latitudes) should be anticipated within the next 45 years.

Before the warming effect of the other trace gases was

appreciated, it was thought that carbon dioxide alone would not generate such a rise until much later—the last half of the next century. The effect, however, of the other trace gases which are increasing rapidly, is to advance the date of significant warming considerably.

The great concern, of course, is that a global warming of 1.5-4.5°C would cause the sea level to rise from 25-145 cm, mostly as the result of thermal expansion of sea water. This would inundate low-lying coastal cities and agricultural areas, and many countries could expect their economic, social and political structures to be severely affected.

As rising seas invade and shift the boundaries of coastal states and modify the levels of shared bays, estuaries and international waterways, local agony could translate swiftly into international crisis. This would be accentuated by the effects of changing climate on inland crops, forests and ecosystems. Although knowledge of these effects cannot be certain until they occur, experts believe that crop boundaries will move to higher latitudes, and will do so much more quickly than forest boundaries because of the longevity of trees. The effects of warmer oceans or marine ecosystems on fisheries and food chains is also virtually unknown.

Melting of small glaciers would result in a further rise of sea level of more than 20 cm over a century. It is now thought unlikely that a global temperature rise of 4.5°C would eventually cause the Western Antarctic Ice Sheet to collapse. But if this occurred, it would add gradually another five metres to ocean levels, it is thought, perhaps over several hundred years. By the time the effects of the rise in sea level begin to be felt, however, the room for preventive and even corrective action will be much reduced, and the situation will become increasingly irreversible. The agony of low lying coastal areas will sharpen sources of local and international tension.

Some policy directions
Nobody knows how communities and nations will respond to these situations as they evolve. Governments, however, and the world community are in a position to anticipate them and to take certain measures to prevent them, reduce their impact or, with climatic change, at least buy time to facilitate adaptation to the consequences.

We seem to be entering an era where sooner or later nations will have to formulate and agree upon long-term policies for all energy-related activities affecting sustainable development and influencing the climate on earth. Given the complexities of international negotiations on such issues and the time lags involved it is urgent that the process start now.

What is, in fact, needed for these energy-related issues, marked as they are by varying degrees of uncertainty, is a three-track strategy. This would combine, firstly, improved monitoring and assessment of the evolving phenomena, secondly, increased research to improve our knowledge about the sources and effects of the phenomena, and, thirdly, the development and implementation of clusters of new or modified economic, financial, trade and sectoral policies that would prevent or reduce avoidable and destructive effects of these phenomena on human health, resources and ecosystems, especially those effects which involve high risks of irreversibility and transfers across generations.

No single nation has the institutional, professional and resource capacity needed to undertake the research, monitoring and assessment now required. And no nation has the political reach to entertain the changes in the structure of policies that will be necessary. Increasingly sophisticated forms of international burden-sharing to address these issues are required at both regional and global levels. This should include an immediate start on an internationally binding convention covering all energy-related environmental and developmental questions.

While this proceeds, more immediate policy measures can and should be adopted. The most urgent, as mentioned earlier, is an effective arrangement to stabilize wild fluctuations in world energy prices and hold them at levels that will maintain steady gains in energy efficiency and a shift in the energy mixture towards more renewables. A wide range of specific measures is available to increase energy efficiency in housing, industry, transportation, agriculture and other sectors. Most measures are very cost-effective and the potential has scarcely been touched. Carbon dioxide output globally could be halved by energy efficiency measures over the next 50 years or so without any reduction of the tempo of economic growth. And, as noted earlier, these measures would also serve to reduce other emissions and thus reduce acidification and air pollution. In addition, they would improve the balance of payments of nations, reduce debt burdens and provide rich sources of revenue for countries to improve fiscal management.

Many other mutually reinforcing measures should also be entertained, for example, measures to deliberately change the fossil fuel mixture. The current global energy mixture (per cent) is oil 41; coal 24; gas 17; other 18. In producing this amount of energy, gas, oil and coal produce carbon dioxide in the ratio of 43: 62: 75 respectively.

Apart from fuel cleaning and fuel switching, financial mechanisms should be established that build the external costs of different energy sources into their price. Strengthened economic incentives and disincentives favouring environmentally attractive energy investments are needed as are emission limits, with licences to reflect them, flexible trading in such licences, and improved regulatory measures.

Some measures would serve to control or reduce other problems. Nitrogen oxides, for example, could be reduced by more careful nitrate fertilizer controls. This may also help to reduce the chronic problem of eutrophication and the growing problem of nitrates in drinking water, which is currently an important health hazard to small babies in many areas worldwide.

Gases other than carbon dioxide are thought to be responsible for about one third of present global warming and it is estimated that they will cause about half the problem around 2030. Chlorofluorocarbons alone are thought to be responsible for 20 per cent of the present day temperature rise, and since they are produced by one large manufacturer worldwide, they might be an obvious starting point for management. A roughly four-fold increase in chlorofluorocarbons is expected before 2050 at present rates of emission; it is estimated that this will cause a reduction in the stratospheric ozone shield of 10-30 per cent, thus allowing greater ultraviolet radiation to reach the earth's surface.

Nuclear energy could also have an important role to play in attacking all fossil-fuel related issues. The conditions under which its development can be maximized are discussed below. They need to be established and implemented through a new and binding international agreement.

Much of the uncertainty concerning the future global energy path could be cleared away if negotiated agreements could set *ceiling* levels for the quantities of the principal transboundary air pollutants (including carbon dioxide) and work backward to map out exactly what energy strategies would be needed in future to peg them below these ceilings. The available range of strategies can provide ample room for national priorities and for the energy supply conditions unique to each nation, but much policy development work is needed to obtain such strategies. This should proceed hand in hand with accelerated research to reduce remaining scientific uncertainties.

What is needed urgently is a concerted act of political will. The uncertainty about the conditions we shall finally reach owes more to policy irresolution at the political level than to any shortcoming in the science of these issues.

Increasing Renewable Energy

The potential for renewables

Renewable energy sources provide more than 20 per cent of the energy consumed worldwide; of this, biomass accounts for 14 per cent and hydro-power for six per cent. Worldwide reliance on these sources has been growing by more than 10 per cent a year since the late 1970s, mainly because of a surge in hydro-power and wood fuels.

Although wood ranks only fourth in the world's energy budget after coal, oil and natural gas, about half the world's population rely on it, mainly for cooking, and its use has been expanding at two per cent annually. Wood fuels are the most important energy source for over two thousand million people in developing countries, where from 30 to 98 per cent of all energy consumed comes from biomass. Although biomass is in adequate supply in many countries, 1,300 million people live in areas that can satisfy minimum needs only through unsustainable overcutting; and over 110 million people live in areas where even with such overcutting minimum needs cannot be satisfied. By the turn of the century, unless counter measures are taken now, three thousand million people will suffer from a severe or absolute scarcity of fuelwood.

Where fuelwood is scarce people turn to agricultural and animal wastes for fuel. The energy potential of such wastes is very high indeed, especially when used in modern conversion devices, such as gasifiers. In most rural societies, however, much of these wastes has other uses, and its diversion for energy purposes may result in problems, particularly the diversion of nutrients from agricultural lands (See Figure 5). Policies for enhanced use of biomass wastes, therefore, should be based on careful local studies.

In contrast, there is a vast unrealized potential for the use of fuelwood in northern countries. In fact, it is becoming a popular fuel for domestic and industrial heating in Canada, USA and the Scandinavian countries.

Hydro-power, which is second to wood among the renewables, has been expanding at nearly three per cent annually. Although hundreds of thousands of megawatts of hydro-power have been harnessed throughout the world, the remaining potential is huge, especially in the Third World.

Solar energy is currently of small importance globally, but it is beginning to assume an important place in the energy consumption patterns of some countries. In many parts of Australia and Greece, for example, solar water and household heating is widespread. Prior to the recent fall in oil prices, Israel expected that 60 per cent of its water would be heated by solar energy by 1990. A number of East European and developing countries have active solar energy programmes and the USA and Japan support solar sales of several hundred million dollars a year. With constantly improving solar thermal and solar electric technologies, it is likely that their contribution will substantially increase.

The Brazilian alcohol programme in 1984 produced about ten thousand million litres of ethanol from sugar cane and replaced about 60 per cent of the petroleum that would have been required in the absence of the programme.[19] The cost has been estimated at $50-60 per barrel of petroleum replaced. When subsidies are removed, and a true exchange rate is used, this is competitive at 1981 oil prices.

Although with the present oil glut the programme has become temporarily uneconomic, it provides additional benefits of rural development, employment generation, increased self-reliance,

PATTERN OF DETERIORATION IN ETHIOPIAN AGROECOSYSTEMS

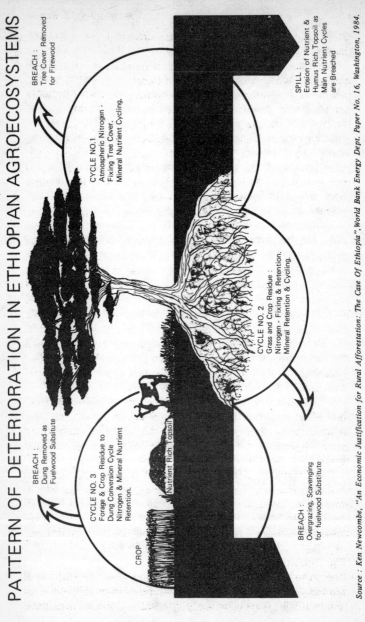

BREACH : Tree Cover Removed for Firewood

CYCLE NO.1
Atmospheric Nitrogen - Fixing Tree Cover, Mineral Nutrient Cycling.

SPILL : Erosion of Nutrient & Humus Rich Topsoil as Main Nutrient Cycles are Breached

BREACH : Dung Removed as Fuelwood Substitute

CYCLE NO. 3
Forage & Crop Residue to Dung Conversion Cycle Nitrogen & Mineral Nutrient Retention.

CYCLE NO. 2
Grass and Crop Residue : Nitrogen - Fixing & Retention. Mineral Retention & Cycling.

BREACH : Overgrazing, Scavenging for fuelwood Substitute

Nutrient Rich Topsoil

CROP

Source : Ken Newcombe, "An Economic Justification for Rural Afforestation: The Case Of Ethiopia", World Bank Energy Dept. Paper No. 16, Washington, 1984.

and reduced vulnerability to future crises in the world oil markets.

The Brazilian ethanol programme is particularly efficient at generating jobs; it requires an investment of $6,000-$28,000 per job, which compares favourably with an average of $42,000 per job for the Brazilian industry sector, and $200,000 per job for the oil-refining, petrochemical complex at Camarcari.[20]

Another promising option is the production of methanol from wood. Like ethanol, methanol has many advantages over petroleum. It produces a clean exhaust gas consisting of carbon dioxide and steam. Some health aspects of using alcohol fuels, however, need careful monitoring. For developing countries, methanol from wood is a viable option only if it is linked to large-scale forestry programmes. An added bonus of such programmes is that it is possible to obtain charcoal as a by-product, a nearly smokeless fuel, able to satisfy a number of end-uses in developing countries, including cooking and artisanal requirements.

The use of geo-thermal energy has been increasing rapidly by more than 15 per cent per year in both industrialized and developing countries. Exploration is expected to uncover a world geo-thermal capacity exceeding 10,000 megawatts by 1990 and the engineering and other experience gained during the past decades could provide the basis for a major expansion in countries rich in geo-thermal resources.

It is difficult to assess the full potential for sustainable development of renewable energies either globally or on a regional or national basis: one estimate placed the long-term maximum potential globally at ten terawatts (See Table 3), a value equal to the 1980 level of energy consumption worldwide.[21]

The potential is clearly very large and with new technologies could expand significantly. While the extent and rate at which renewables are developed will depend on technology, in the short run the policies that attack certain economic, environmental and institutional constraints will be even more important.

The fuelwood crisis

This is especially true for those developing countries with a vast number of poor, rural and urban households, whose livelihoods and well-being depend on access to local supplies of traditional

31

Table 3
World Use of Renewable Energy, 1980, 2000 and Potential (Source: Deudney & Flavin—1983)

Source	1980	2000	Long-term potential
		(exajoules)	
Solar energy: passive design	<0.1	3.5-7	20-30
Solar energy: residential collectors	<0.1	1.7	5-8
Solar energy: industrial collectors	<0.1	2.9	10-20
Solar energy: solar ponds	<0.1	2-4	10-30+
Wood	35	48	100+
Crop residues	6.5	7	—
Animal Dung	2	2	—
Biogas: small digesters	0.1	2-3	4-8
Biogas: feedlots	<0.1	0.2	5+
Urban sewage and solid waste	0.3	1.5	15+
Methanol from wood	<0.1	1.5-3.0	20-30+
Energy crops	0.1	0.6-1.5	15-20+
Hydro-power	19.2	38-48	90+
Windpower	<0.1	1-2	10+
Solar photovoltaics	<0.1	0.1-0.4	20+
Geothermal energy	0.3	1-3	10-20+
Total	63.5	113-135	334-406+

+ indicates that technical advances could allow long-term potential to be much higher, similarly, a range is given where technical uncertainties make a single estimate impossible.

< means less than.

Source: Worldwatch Institute

fuels such as fuelwood, cow dung and crop residues. At the same time unprecedented pressures are being placed on the same biomass base from the agricultural and urban-industrial sectors.[22] Forests are being cleared at a rapid rate to open up new agricultural land and much of this is induced by the need to produce more goods for export (tea. coffee, meat, etc.) to pay debts incurred to import or develop new sources of energy.

Some forest loss is inevitable, of course, to make way for human settlements, agriculture and industry. The continued rapid loss of forest cover, however, will be ecologically, environmentally and, consequently, economically disastrous for the regions concerned. Many of society's most important values are locked up in forests and disappear with them. Additionally, deforestation causes soil degradation and erosion, siltation of reservoirs, reduced electricity production, flooding and loss in agricultural yields.[23]

When fuelwood is no longer available in an area rural people turn to burning biomass wastes, such as cow dung, rice husks and cotton stalks. In some cases these practices do no harm since a true waste product is incinerated (e.g. cotton stalks). In other cases, however, much needed organic nutrients are diverted from the soil (e.g. animal manure, some wet biomass, etc.). Since developing countries are already using a great deal of renewables, mostly in the form of woody biomass (See Figure 6), agricultural and animal wastes, the concern is not necessarily to introduce new systems, but primarily to *modernize* and make more *efficient* existing systems, as is assumed in the low projection.

One fact, however, seems certain. While we cannot go on using fuelwood and other biomass sources the way we do today, no significant substitution seems possible in the near and even medium-term future. Given this immediate, and basic need for domestic fuel, and the low level of substitution possibilities, it seems, at least in the short term, that the only way out of this problem is to treat fuelwood like food, and grow it as a subsistence crop.

Scarcity of fuelwood, however, represents more than just lack of an energy source. In addition to being sources of fuelwood, trees in rural communities are an important source of livelihood for people, In particular, they supply fruit, fodder, fertilizer,

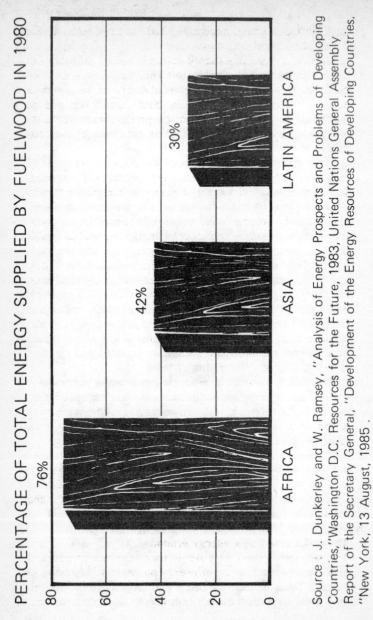

PERCENTAGE OF TOTAL ENERGY SUPPLIED BY FUELWOOD IN 1980

AFRICA 76%

ASIA 42%

LATIN AMERICA 30%

Source : J. Dunkerley and W. Ramsey, "Analysis of Energy Prospects and Problems of Developing Countries,"Washington D.C. Resources for the Future, 1983, United Nations General Assembly Report of the Secretary General, "Development of the Energy Resources of Developing Countries. "New York, 13 August, 1985 .

medicines, poles for construction, shade and raw materials for a host of artisanal activities.

Wood as a renewable energy source is usually thought of in the context of the traditional sector. This, however, is far from the whole truth. Wood is an important feedstock for advanced energy conversion processes in both developing and in industrialized countries, including the production of electricity and potentially of other fuels, such as combustible gases and liquids.

Given present trends, in the next few decades the woodfuel crisis will undoubtedly worsen and the theoretically renewable supply of trees will increasingly become non-renewable. Population will grow substantially, and even if per capita consumption remains steady, total woodfuel demand will grow. Proportionately more of this population increase will take place in urban conglomerations, where generally use of charcoal is higher. Overall, given the inefficient charcoal kilns now widely used, charcoal requires twice as many trees as straight fuelwood to deliver the same energy at the point of end-use. Consequently per capita wood requirements will grow. Finally, urban and industrial demand for wood-based fuels will increase also because of the high price of alternatives, whether in the form of traditional fuelwood and charcoal, or as feedstocks for more advanced biomass conversion systems.

The challenge ahead for developing countries is tremendous: to stabilize, and increase sustainable wood production to satisfy all legitimate demands in all sectors, at a time when unprecedented and increasing pressures are exerted on these same resources from agriculture and industry.

Often the picture is further complicated by complex socio-political factors. In many areas, in spite of a relative abundance of trees, the availability of woodfuels is very low. The reasons include problems such as access to and ownership of trees, and the role of women in society.[24]

Toward a renewable energy economy

The policies required for a steady transition to a broader and more sustainable mixture of energy sources are beginning to emerge. Those associated with the high scenario, however, with their emphasis on fossil fuels and other conventional sources, would have a negative effect on this essential process. National and international institutions would remain attached to supply-side planning and the promotion of conventional fuels. The limited budgets available for energy research and development would continue to focus almost entirely on improving conventional sources.

In contrast, the policies implicit in the low scenario could favour the promotion of renewables. The low scenario, for example, requires decentralized energy systems, able to satisfy a wide range of end-uses and employing the most economically available sources. Very often, there will be various forms of energy saving measures: insulation for space heating and cooling; fuel-saving vehicles; energy efficient production machinery or processes; co-generation of heat and power; etc. Or, there could be various forms of locally available renewables: active or passive solar; biomass; etc.

With the exception of large-scale hydro-power, most renewable energy sources are compatible with the less intensive, decentralized systems that now exist in most developing countries and that will emerge in other parts of the world. In fact, most renewables are characterized not only by decentralization, but also by diversity. Unlike non-renewables, this enables each country to tailor its energy economy to its indigenous resources. Although every country possesses some sun, wind and biomass, some are better endowed with certain renewables than with others. Some may wish to rely on one locally abundant form of renewable energy; hydro-power, for example. Others may be able to develop highly diversified renewable energy economies. Biomass gasifiers, biogas plants, small hydro-power installations, and most solar and wind systems work well on a small to medium scale and are often best suited for family and village or community use, especially in developing countries. Some large countries or regions, for example Brazil, India or North America, may be wise to develop the entire panoply of renewable sources from energy crops, including alcohol and fuelwood plantations to wind, solar and photovoltaics.

As they evolve, renewable energy economies will also reinforce the self-reliance goals of most developing countries. Certain renewable energy systems could either replace imported fuels, or satisfy new demands without the need to import additional energy. This is particularly true of hydro-electric plants, biomass gasifiers, biogas plants, ethanol distilleries, various solar electric and solar heat systems.

For developing countries, reliance on locally available energy sources should improve balance of payments and reduce debt loads. The outlays of foreign exchange are negligible, particularly if conversion hardware is manufactured locally. In addition, many renewable sources are virtually inflation-proof. Once the initial power systems is made, the cost of running hydro and solar power stations is tied to maintenance.

Some policy directions
A number of policy measures are essential to speed the transition to renewables, policies that reinforce the low energy scenario and are reinforced by it.

The economic, social and environmental consequences of using renewables require serious evaluation and comparison with non-renewables. Sustainability is the limiting factor with some renewables. Using them above sustainable limits will render them non-renewable and consequently unsustainable. It is very important that these limits be continuously assessed and monitored in every country. Countries should undertake both a full assessment of their potential for renewable energy, and an assessment of the economic, social and environmental consequences of a steady shift to renewables.

The influence of unreasonably low fossil fuel prices has already been noted. The high level of hidden subsidies for conventional fuels built into the legislation and energy programmes of most countries also distort choices against renewables. These subsidies are legion, including not only research and development, but also depletion allowances, tax write-offs and direct support of consumer prices. Countries should undertake a full examination of all subsidies and other forms of support given to various sources of energy and should publish the results.

Beyond that, given the advantages to their development of a renewable energy economy, countries should also reorient certain fiscal and tax policies that now penalize renewables. At a minimum, such policies should be neutralized and in some cases recast to give individuals, entrepreneurs, private com-

37

panies and parastatals the promise of sufficient gain to induce them to invest in new energy technologies which improve both economic efficiency and the environment. Developing countries should consider the establishment of small venture-capital funds at national, regional or global level to finance renewable energy projects. International development assistance agencies and financial institutions should assist in this.

Both industrialized and developing countries should evaluate critically the feasibility of changing the basis of thermal power generation from the combustion of fossil fuels to gasification. Where available, biomass fuels should be included in the assessment. Following comprehensive economic analyses, wherever feasible, countries should step up the pace of change from coal and oil to gas and renewable energy sources, such as solar thermal, solar photovoltaic, biomass, and eventually, subject to the resolution of the waste disposal, safety and non-proliferation issues, to nuclear power. In particular, the transport fuel base should be shifted from petroleum and diesel to alcohol fuels where a full socio-economic comparative analysis shows this to be worthwhile under local market conditions.

Progress in renewable energy technologies depends on a much greater level of research and development. The current effort has registered some success, but in most countries is dwarfed by the effort put into fossil fuels and nuclear power. In addition to increasing the level of direct support, all countries should offer high tax write-offs to corporations that undertake research into new energy technologies or invest in energy-saving or new energy technologies.

The change to a renewable energy economy in developing countries must include strategies for the stabilization and increase of sustainable wood production to satisfy growing demand, both in the traditional and modern sectors. This has to be carried out in environmentally favourable ways, which enhance and do not reduce the countries' agricultural priorities.

In particular, existing trees and forests will have to be managed in environmentally and economically sustainable ways. Additional trees and forests should be planted in different ways; agroforestry programmes, large- or small-scale dedicated energy plantations, village- and community-scale woodlots or as conventional forests. The aim should be to make tree growing financially viable. Where possible, wood conversion processes should be made more efficient, particularly in charcoal making.

In the subsistence sectors, the need is to make both food *and* fuelwood available where there is a local demand. The production of food and fuelwood in distant regions and forests does not help the rural poor. The definition of what is local is site-specific, but presumably extends to that boundary within which distribution is still feasible.

Particular attention should be paid to enhancing fuelwood supplies for subsistence farmers (where most fuelwood is still collected as a *free* good). Extension work, appropriate agricultural and agro-forestry programmes, mostly organized by local communities and NGOs and, in particular, the role of women, must be strengthened and supported.

In examining their current structures of energy subsidies, the governments concerned should, therefore, pay special attention to the fuelwood crisis. Those fiscal, tax and other policies that induce practices which add to pressures on fuelwood should be reversed and special programmes of incentives to farmers to augment fuelwood resources should be introduced. These could include annual loans to farmers based on the number of trees surviving every year, until the initial crop is harvested; these loans could be recovered from the sale of the crop.

Institutional barriers to renewables are formidable in many countries. Electrical utilities, for example, have often enjoyed a complete monopoly not only in power distribution, which is normally justifiable, but also in power generation. In some countries, a relaxation of these powers, so that utilities are required to buy power generated by industry, small systems and individuals at avoided costs, has created many opportunities for the development of renewables. Beyond that, requiring utilities to adopt an end-use approach to planning, financing, developing and marketing energy can open the door to a wide range of energy-saving measures as well as renewables. All countries should consider this urgently and international development agencies should support it.

More efficient woodstove programmes help in saving wood, but their main benefits will be cleaner and healthier kitchens and more available time for women to carry out activities other than searching for fuel. More efficient stoves and other wood-saving technologies, such as aluminium pots and pressure cookers, make sense in areas where wood is purchased, and saved wood translates into saved money. In rural areas, changing cooking habits is more difficult, and from an energy point of view not strictly necessary. General improvements in the standard of living as the result of increased agricultural or small-scale industrial production, etc., often result in people opting for more efficient cooking and other energy conversion appliances. Clearly, countries should adopt all possible methods of increasing the efficient use of renewables in the home. Where efficient distribution can be assured, developing countries should develop a system to gasify biomass used as cooking fuel. Developing countries should also exploit crop residues as an important source of biomass and use them scientifically to yield cooking fuels.

Making Nuclear Energy Acceptable

Nuclear energy is one of the most important potential sources for the centralized production of electricity for the future.[25] It is part of the quest for clean power in the post-fossil-fuel world, primarily of interest in industrialized countries, although some of the newly industrialized countries which are not well endowed with other resources also have ambitious programmes. Both the *high* and the *low* projections assume a certain growth of nuclear power. In the former, the projected installed capacity by 2020 would be 8,100 gigawatts, in the latter, 800 gigawatts—compared with 200 gigawatts installed in 1980.

Nuclear energy, however, is facing a number of serious problems. The resolution of the problems of radioactive waste disposal, including the decommissioning of nuclear reactors, and the public perception of the nuclear fuel cycle are the most important, although other critical issues exist, such as safety questions, trade in nuclear technology, the appearance of new types of reactors and the proliferation of nuclear weapons.

The Chernobyl accident strengthens our view that no nation can make decisions on energy matters on its own. Just as the fossil-fuel emissions from a plant may cause harm in a neighbouring or even in a distant country, or the damming of a river may have implications in a neighbouring country, nuclear safety issues also have regional and even global implications. A sovereign nation may decide that lower safety levels can be tolerated in a nuclear reactor, but the radioactive fallout released in an accident does not respect national boundaries. Nuclear safety is a regional, often a global concern, and consequently should be the subject of internationally binding agreements and conventions supervised by the International Atomic Energy Agency (IAEA) and other international bodies.

Such an agreement should provide that, if an accident occurs, neighbouring countries would be notified immediately and automatically. Detailed information about the nature of the accident should follow quickly; such information should concern not only the release of radioactive materials, but also any other matters necessary to enable neighbouring countries to take appropriate preventive action.

In most countries nuclear power agencies have lost the confidence of the public. There seems to be a mixture of various strategies which would have to be followed for a long time to regain this confidence.[26] These strategies will have to include a demonstrated record of safe operation of the nuclear fuel cycle, the installation of new reactor types which are inherently safe rather than inherently unsafe, much more openness in all sectors of the nuclear fuel cycle, and resolution of the outstanding questions of the decommissioning of nuclear reactors and radioactive waste management, including sites for the permanent disposal of high-level radioactive wastes. The clear separation of civilian and military nuclear activities in all countries is also necessary.

The resolution of the problem of disposal of radioactive wastes will not be easy, partly owing to the extreme complexity of any of the proposed or potential disposal methods, but mostly to the fact that *absolute* proof of safe disposal is impossible; only time will tell if the disposal configuration chosen today will function as designed. This time can be as long as a million years. It must also be kept in mind that, while very little data on the sources of radioactive waste is available from industrialized countries, which are members of the nuclear club, a large proportion of those wastes come from the processing of nuclear fuels for non-pacific purposes, i.e. the production of nuclear weapons.

Some countries have introduced legislation tying any further growth of nuclear energy and the export/import of nuclear reactor technology to a satisfactory solution of the problem of disposal of radioactive wastes. This has intensified the search for environmentally acceptable forms of management of these wastes and has brought some nations closer to solutions for their final disposal. We welcome this, and urge that all nuclear countries introduce similar, and increasingly strict legislation.

As a result, the technology of disposing of even long-lived, high-level wastes in environmentally acceptable ways has now reached the stage of technical feasibility, and a number of nations are well under way in the implementation of waste disposal programmes.[27]

The extremely long time horizons create particularly difficult problems. The question arises whether this generation has the right to bury in geological formations wastes which may harm future generations, however small that probability may be. In fact, geological disposal of long-lived radioactive wastes is an attempt to bypass the polluter pays principle by transferring some of the potential costs and radiation harm of the nuclear fuel cycle to generations not yet born. This is more than a theoretical question, since, at least in principle, it is possible to transmute long-lived nuclides to ones with much shorter half lives, although at higher cost.[28]

We propose that all governments establish a goal of working toward *closing* energy cycles. The aim would be to extend the polluter pays principle into the time dimension. The ideal of achieving clean-up at exactly the same time as energy production will in most cases be impossible, but efforts should be made to reduce the time delay as much as practically possible, preferably to within the same generation. In particular, we recommend that the *transmutation option* for the elimination of long-lived radioactive wastes be further studied and developed to reach the stage of technical feasibility in order to allow full cost/benefit comparison with the *geological disposal option* now practised.

The disposal of radioactive wastes is one area which lends itself well to international co-operation. Small nations do not possess all the scientific, technical capability to carry out the necessary research and development work. Some nations have superior/inferior geologies for the disposal of radioactive wastes. Consequently, the scope for international co-operation in the development of waste disposal methods, and where possible in international/regional repositories, is great.

In this context, there is urgent need for the creation of an international review group, consisting of top-level experts, independent of their governments, to generate guidelines and evaluate national proposals for the disposal of radioactive wastes—a role similar to that of the International Commission on Radiological Protection (ICRP) in the area of radiation dose and protection.

Recently, some industrialized nuclear countries (e.g. Federal Republic of Germany, Switzerland) have shown interest in making agreements with developing countries (e.g. China, Sudan, Ethiopia, etc.) whereby the latter would take the nuclear wastes of the former for a fee. Although at the time of writing no such agreements have been signed, the possibility is very dangerous indeed. We consider this irresponsible, even when the recipient country has agreed to do so. Such a practice would be yet another attempt to bypass the polluter pays principle by shifting some of the potential costs and radiation harm of the nuclear fuel cycle away from where the beneficiaries are located to a different geographical location. Radioactive wastes should be disposed of in the territory of the country enjoying the benefits of nuclear power. On the rare occasion when this is not possible for geological or other reasons[29], disposal in another country should be subject to strict international safeguards, supervised by the IAEA and the independent review group proposed above.

It is likely that in the future a number of new countries will join the international nuclear reactor buyers. One important area where action will be required is an international agreement on suppliers and receivers concerning the transfer of nuclear technology. The agreement would have to cover assurances for the disposal of all resulting radioactive wastes, safety standards and licensing procedures.

The nuclear reactor industry is now going through a tremendous change. A host of entirely new types of reactors, including relatively small ones (10-300 megawatts, compared to the typical 1000 megawatts of the present generation of reactors) are now under development, many of which are designed for potential markets in developing countries.[30] Some industries are planning 100% package deals, in which the reactor could be assembled at the factory, shipped to location on rail or barges, and hooked into a national grid.

The problems as well as the potentials of these developments are tremendous. On the one hand, standardized construction will enable manufacturers to ensure higher safety. On the other hand, problems such as technological dependence, as well as the need for highly skilled operators will have to be addressed by the receiver countries.

Increasing Energy Efficiency

Energy efficiency should be the cutting edge of national energy policies and sustainable development; and measures to achieve it deserve the highest priority on national agendas. Although impressive gains in energy efficiency have been made since the first oil shock, as seen in Figure 3 above, the results have been uneven. Some countries have made great progress, others have gone backwards. The latter's potential is not limited to attaining the achievements of the former; for both the opportunities for future gains are enormous. But both will need to employ deliberate policies.

The cost-effectiveness of efficiency as the most environmentally benign source of energy is well established. There are many examples where energy consumption per unit of output for *best practice* technologies is a half or less than half of typically available equipment. This is true of lighting, refrigeration and space cooling, which are growing rapidly in most developing countries and putting severe pressures on electricity supply systems. It is true of cooking fires and cooking equipment, with all their impact on tree cover, of the recycling of crop and animal residues now burnt for fuel, and of soil erosion. It is true of cultivation and irrigation systems, of the automobile, and of many industrial processes and equipment. With many of these, the time required to pay back the additional investment in energy savings is often less than two years and is frequently measured in months or weeks.

These claims are often rejected by developing countries, and the poor generally, as concerns of only the extravagant and well-to-do. Nothing could more grievously misrepresent the truth. It is the poorest who are most often condemned to use energy and other resources least efficiently and productively, and who can least afford to do so.

The woman who cooks in an earthen pot over an open fire uses perhaps eight times more fuel than her affluent neighbour with a gas stove and aluminium pans. The poor who light their homes with a wick dipped in a jar of kerosene get one hundredth of the illumination of a 100-watt electric bulb and use just as much energy to do so. The cement factory, automobile or idling irrigation pump in a poor country is no different from its equivalent in the rich world. In both there is roughly the same scope for reducing the energy consumption (or peak power demand) of these devices without loss of output or welfare. But in a poor country the benefits thus gained will mean much more.

This, however, is the tragic paradox of poverty. It is not energy, but rather poverty which is the limiting factor for the poor. They are forced to live on a meagre current account, and thus use inefficient equipment, because they have no savings for investment capital to purchase energy-efficient devices. Consequently, they end up paying many times over for a unit of delivered services.

While there are countless examples of successful energy efficiency programmes in industrialized countries, similar programmes face a large number of barriers in developing countries. In all countries, ignorance tends to be the most important constraint. Many consumers, including large industries, do not know how they use energy, what it costs them, how costs can be reduced, or how to set about reducing them. Information campaigns in the media, technical press, schools, etc.; demonstrations of successful practices and technologies; free energy audits; energy *labelling* of appliances; training in energy-saving techniques; and other techniques have been used successfully to increase awareness. They need to be extended.

Indifference is equally widespread. Apart from transportation and the most energy-intensive industries, energy typically accounts for only three to five per cent of total costs of an enterprise. The figure is much the same for middle-to-high income

45

families in industrialized countries. It represents only one to three per cent of family budgets in warmer developing countries where there is little need for space heating. Where budgets are tight, therefore, investments in energy savings may be postponed regardless of the potential pay-off.

This tendency is reinforced by energy pricing policies, which may reflect subsidies and almost never reflect the real costs of producing or importing the energy, including those of external health, property and environmental damage. Countries should evaluate the overall cost to government and society of different energy options, both renewable and non-renewable, with all hidden and overt subsidies included. Economic pricing of energy, perhaps with safeguards for the poor, needs to be extended in all countries.

Developing countries face particular constraints in this area. They are generally in the throes of foreign exchange conditions which make it difficult to purchase efficient, but costly, energy-conversion devices. Often such technologies are new, not yet fully tested, and poor developing countries cannot take the risks of experimenting, and possibly failing, with such technologies. Finally, many measures for energy savings tend to be the *fine tuning* of already functioning systems. These do not appear as attractive to aid agencies or to local government officials as new, large-scale energy-supply projects.

There are more subtle but no less important price and cost distortions. The economic benefits of improved energy efficiency, for example, may be acquired by parties other than the industry or consumer who has to bear the investment costs. Energy efficiency measures which reduce peak electricity demand and thus postpone the need for investment in additional capacity are an example. Frequently, the ratio of the cost of avoidable supply to the cost of the efficiency measure is two or three to one. In these and similar cases, there are strong arguments for systems to enable those who invest in energy-efficiency measures to receive more of the financial rewards.

Many energy-efficiency measures cost nothing to implement. Where investments are needed, they are frequently the main barrier to action, even when pay-back times are short. These barriers are often absolute for the poor consumer or for small informal sector entrepreneurs. For the latter, subsidies will be necessary. Where investment costs are not insurmountable, there are many mechanisms for reducing or spreading initial investment which can be adopted by the public or private sector; these range from subsidies, tax credits and loans to *invisible* measures such as payment through reduced energy bills.

Mandatory efficiency standards for equipment and appliances have four great advantages over strictly market measures. They can be relied upon to produce predictable savings in energy, thus greatly assisting energy supply planning which must often be based on expected demand some 10-15 years ahead. They overcome the reluctance of manufacturers to take a gamble on producing more energy-efficient technologies for uncertain consumer markets. They can induce both the spread of existing and the innovation of new technologies, processes and products, as demonstrated by automobile, chemical, steel and other industries. They can induce and stimulate research and development efforts towards socially desirable goals.

The imposition of minimum-energy consumption standards on the manufacture, import or sale of equipment is one of the most powerful and effective tools in promoting energy efficiency. Where the equipment concerned is traded internationally, these may require international action. Countries, and where appropriate, regional organizations, should introduce and/or extend increasingly strict minimum efficiency standards for equipment and mandatory labelling of appliances.

Transport

Transport has a particularly important place in national energy and development planning. It is a major consumer of oil and accounts for 50-60 per cent of total petroleum use in the majority of developed and developing countries. It is a major source of local air pollution and regional acidification of the environment.

Looking to the year 2000 and beyond, vehicle markets will grow much more rapidly in developing countries, adding greatly to potential air pollution in cities where international norms are already exceeded. Indeed, unless strong action is taken, air pollution could become a major factor limiting industrial development in many Third World cities. In this context, fuel economy emerges as the most cost-effective means of both preventing further growth in air pollution from vehicle transport

47

and of preserving a region's capacity for sustainable development.

In the western industrialized countries, energy use in transport grew quite slowly after 1973 and began to decline in the late 1970s. Higher fuel prices and strong competitive forces, leading to large and rapid increases in energy efficiency of the air traffic and automobile fleets, were mainly responsible for this reversal. Progress was uneven, however, and there is a wide difference between the most and least efficient automobile fleets. The Japanese automobile industry led the way and its fleet is the most efficient with an average of eight litres per 100 kilometres. The North American industry brought up the rear and at present averages a little below 16 litres per 100 kilometres.

With higher prices, fuel economy becomes a high-visibility issue for consumers as well as governments. It can continue to be a driving force behind technical innovations directed at dealing with a changing operating environment and at gaining competitive advantage in the market place. In the absence of higher prices, however, mandatory standards providing for a steady increase in fuel economy may be necessary. Either way, the potential for substantial future gains in fuel economy is enormous; improved body design, substitution of materials, and engines and power trains are some of the technical paths now being pursued. If the momentum can be maintained, the current average fuel consumption of approximately ten litres per 100 kilometres in the fleet of vehicles in use in developed countries could be cut in half by the turn of the century.[31]

A key issue is how developing countries can secure similar improvements in the fuel efficiency of their fleets. Those countries that import their fleet could mandate standards for new vehicles. In those countries where vehicles are assembled under license from manufacturers in an industrialized country, however, the situation is different. The designs are frequently old fashioned and predate energy-efficiency improvements. These countries should give priority to the reform of licensing and import agreements so that they will have access to the best available fuel-efficient designs and production processes.

The issue goes deeper, however. Transport is the one sector of demand where there are at present few substitutes for oil.

While the scope for increases in energy efficiency in road haulage is immense, and will lead to very significant savings in energy, fleet life in developing countries is nearly twice that in industrialized countries. Thus the rate of improvement in energy efficiency will be much slower, and will almost certainly fall behind the rate of growth in total transport demand, at least if historical trends in the latter are any indication of what will happen in the future. Thus, even under the most optimistic assumptions, developing countries could face a significant rise in the total demand for transport fuels over the next three to four decades.

In view of this, certain developing countries should explore the potential of non-oil-based transport fuels. Some actually have done so, Brazil being the outstanding example. The obvious candidates are alcohol-based fuels: ethanol which is already in use in many developing countries and methanol which has so far been used mainly in racing cars because of its superior flame speed. Both ethanol and methanol can be obtained from biomass, and methanol may also be obtained from coal and natural gas. In principle, their combustion yields only steam and carbon dioxide, but some recent research and experience indicates the possibility of other health hazards. As a result, there is an urgent need to monitor the use of ethanol and methanol closely, and to conduct research on the elimination of potential hazards similar to that which has been done over a long period for hydrocarbon fuels.

The encouragement of more energy efficient modes of transport is another option. In developing countries—and historically in the industrialized countries—the trend has been the other way. In particular, rail has given way to road and the bus to private vehicles. Reversing this trend is not easy, since the least efficient transport modes are the most convenient and flexible. And the provision of quality services that attract customers, in particular, more frequent and less crowded buses or trains, inevitably reduces overall fleet energy efficiency.

Changes of this kind are easiest if they are built in to development strategies. If city growth is centred on multiple cores, or on new towns within an expanding metropolitan region, efficient and rapid mass transit systems can be used to link the centres in which most travel can be non-mechanized.

Such major shifts in development patterns can often be achieved by simple legislative measures such as zoning and floor-space regulations.

Industry

Industry is also a major source of energy demand and accounts for 40-60 per cent of all energy consumed in industrialized countries and 10-40 per cent in developing countries. Like transport, it is a major source of pollution, especially in those countries that have not enjoyed strong environmental programmes over the past two decades.

Most trends point to a very rapid growth of industry by the turn of the century, but the form and pattern could be markedly different between industrialized and developing countries. Industry in the former has been undergoing a massive restructuring marked by a shift toward higher technologies, a substitution of synthetics for primary inputs, and a growing dematerialization of the economy.

At the same time, there has been a significant improvement in the energy efficiency of production equipment, processes and products. Introduction of these improvements has been stimulated, in part, by higher energy prices. As a result, in every sector of industry there are now plants that are comparatively energy efficient and hence environmentally efficient and economically competitive. This is true of those sectors most frequently associated with newly industrializing countries; these include iron and steel, non-ferrous metals, bulk chemicals, pulp and paper and food processing. These technologies should spread to older plants as capital equipment is replaced.

During this period, there has also been a significant shift of industrial capacity in the basic and traditional sectors of developing countries to industries that tend to be more energy intensive and polluting. Moreover, developing countries have tended to attract older technologies and processes which are comparatively inefficient and internationally non-competitive. While such technologies may be foisted on unsuspecting or unaware governments in some developing countries, it must be admitted that some developing countries actively seek to buy old plants as part of their industrial strategy. If these trends continue, developing countries could end up with industrial sectors that

50

are not only economically uncompetitive but also highly polluting and which impose heavy health, property and environmental-damage costs on their cities and economies.

In this context, energy efficiency again emerges as the most cost-effective means of both preventing the further growth in such damage costs and promoting a competitive and sustainable industrial sector.

The key issue, again, is how developing countries can ensure that future industrialization reflects the most advanced and resource-efficient technologies available in each of the sectors concerned. Several measures seem within reach. Those countries that permit nationals or parastatals to import plants on a turn-key or other basis, or permit multinationals to establish plants on their territories, should ensure that all permits, licences and contracts provide for the best available energy and environmentally efficient technologies and processes. Moreover, such contracts should require registration of complete plans for the safe management and disposal of all emissions inside and outside the plants, and of all wastes. Development assistance, export credit and other international financing agencies involved should ensure that these are included in the financial plans of the industry.

Developing countries often need to decide on the comparative advantages of domestic production of industrial components. In these cases, the most energy intensive components might be imported, and the others made domestically, thus achieving a far lower overall energy intensity for the final product. Energy saving of as much as 20-30 per cent could be achieved by such skilful forms of industrial development.

The proper maintenance of industrial plant, especially older equipment, can also save much *down-time* and pay real dividends in terms of energy saving. Industry-oriented energy conservation programmes, managed perhaps by an *energy service utility* with incentives to help existing industries to identify cost-effective opportunities for saving energy, could reduce energy demands by a further one third. Savings of this order can not only improve the competitiveness of a nation's industrial sector, but also improve its balance of payments, reduce its debt requirements and render the environment in Third World cities capable of accepting more development.

51

Where the latest industrial designs are transferred to developing countries, either through retro-fitted process improvements or through new plant designs, the improvement in energy efficiency can be truly dramatic. This improvement is vitiated only by the much longer life of plant and equipment in developing countries. This longer life stems only partly from the scarcity of capital and foreign exchange. An equally important reason is the existence of protected home markets and the monopolistic structure of production which make improvement unnecessary for raising or retaining high levels of profits. This means that even a modest package of fiscal and monetary incentives can significantly accelerate the pace of technological renewal, even within the given capital constraints.

Agriculture

Globally, agriculture is only a modest energy consumer, accounting for about 3.5 per cent of commercial energy use in industrialized countries and 4.5 per cent in developing countries as a whole. A strategy to double food production in Third World countries through massive increases in fertilizers, irrigation and mechanization would add only 140 million tons of oil-equivalent to their agricultural energy use. This is only some five per cent of present world energy consumption and is almost certainly a small fraction of the energy that could be saved in other economic sectors in the developing world through appropriate energy efficiency measures.

Agriculture is usually the least energy-intensive sector in national economies and the one with the highest economic and social return for each extra unit of energy input. The western industrialized countries have established clearly that the *high food-high energy* linkage can be broken. While energy use has grown, energy efficiency has grown even faster, permitting a significant rise in productivity.

Much of the increase in efficiency is due to the use of chemical fertilizers and pesticides. Failure to manage them properly, however, is now threatening the sustainability of agriculture in many areas. Increased pest resistance, loss of soil fertility, eutrophication of lakes, contamination of streams and nitrate pollution of water supplies are some of the effects undermining the potential for future gains in production at the rates enjoyed in the past. Measures for more effective management are dealt with in another report to the Commission.

Agriculture in developing countries, on the other hand, suffers from low levels of energy use and productivity; the potential for increasing both is thus enormous. It is hard to find examples where increasing levels of energy use do not bring more than proportionate increases in yield, income and profits.

Fertilizers are perhaps the most significant example. Perhaps the most important means available to developing countries to secure the annual gains in food production needed for their growing populations is to increase yields through higher applications of appropriate nutrients. In this regard, locally available sources of organic fertilizer can be more fully exploited. Some 10-15 million tons of nitrogen and five million tons each of potassium and potash could be found in the Third World if only half the available human and animal manure were used. A comprehensive composting or biogas programme in the Third World could provide an estimated 50-100 million new jobs and in the latter case produce high-grade energy for cooking, lighting and irrigation pumping.

Selective mechanization with small machines and improved use of animal draft power are also important examples of energy intervention to break labour bottlenecks, they significantly improve productivity and, in many cases, allow double or even triple cropping. Access to more conventional sources of power would also pay high dividends in increased productivity. Farmers, for example, require energy to pump water for irrigation and other uses, or diesel for tractors, and they require it at precise times of the year. If they do not get it because of priorities elsewhere, yields suffer or crops may fail entirely.

The main constraints on increasing energy for agriculture in developing countries can be traced to unbalanced and inequitable development policies. Although there are vast differences in the political and economic power of rural societies to command energy resources, and genuine problems of resource distribution in rural areas, a balanced development strategy could achieve much in minimizing these problems.

3. Institutional Requirements of the Transition

Attention was drawn at the beginning of this Report to the ways in which the oil price rises of 1973-74 and 1979 actively stimulated a movement away from oil into other forms of energy and towards a concern for saving and efficient usage practices. Many countries began to review their national energy polices and there was great interest in coal, nuclear power, natural gas and renewable forms of energy. In the early 1980s, after the initial economic upheavals had settled, many began to see the steady transition from oil to other forms of energy as a healthy sign in a world that had formerly been too heavily dependent on oil products. At that time, predictions and forecasts about the future of oil abounded, most of them expressing pessimism about future prices. The few analysts who prophesied drastic falls in oil prices as producers competed for a share in a shrinking world market were quickly dismissed. Nowadays, with prices dropping almost as dramatically as they originally rose and in real terms once more approaching the 1973 level, most serious analysts have a much healthier respect for uncertainty and many feel that the journey to any energy future involving oil may well mean a very bumpy ride for consumers and producers alike, unless radical action is taken now.

In spite of this, energy seems to have slipped off the global agenda at the moment and many are asking whether worldwide preoccupations with conservation, renewables and coal are really necessary in a world of oil-glut.

Despite this, for reasons connected with more stable and sustainable pricing policies and production arrangements, and for longer term oil availability, the transition must somehow be encouraged to continue.

Already many countries are learning that rapidly falling oil prices can be economically very dangerous. The ground lost towards transition must somehow be regained and more orderly oil pricing and disciplined production restored. Increasing diversity in use of fuel must be revived. This is in everybody's interest. But in the current climate, it is becoming ever clearer that at present the world lacks the economic and financial structures, as well as the institutional arrangements, and, above all, the mental attitudes to do so decisively.

After the price debacles of the last decade or so, many analysts would argue that oil is too important in the global economy to be treated as a mere commodity, competitively traded on volatile world markets. As an alternative both producer and consumer countries should develop policies designed to build some form of limited but appropriately effective convention for the more orderly production and marketing of oil to the year 2000 and beyond.

National Energy Agencies

An important aspect of the transition is the use of alternative energy sources in ways which are environmentally benign and developmentally sustainable but also cost-effective. Despite the immense amount of recent work done on the health/environment risks attached to various energy sources, very little attention has been paid to the true total costs to society of fuel production, transport, energy liberation and satisfactory management of solid, liquid and gaseous wastes. In short, we still lack genuinely comprehensive systems analyses of complete energy cycle costings with the result that integral components of the energy production process, usually waste management, are not included in the pricing of the resultant energy.

This produces serious price distortions between energy sources. Thus coal, or electricity from coal, without waste management costs included, is bound to appear less expensive than the gas option which is cleaner in health/environment terms.

Another important feature is that energy sources vary widely, from wood fuels, crop-wastes, animal dung, through fossil fuels (gas, coal, oil) to electricity produced by hydraulic, nuclear, thermal or solar power. In many countries, responsibility for each of these is placed in separate compartments with little or no contact between them. And when, as has happened recently in many countries, a Ministry of Energy is formed, there are still serious and apparently insoluble demarcation disputes about such issues as whether fuel wood and crop-wastes should be handled by Energy, or be lodged with Agriculture, Forestry or the Environment. It is also often unclear who has the responsibility for managing the environmental impacts of energy production, and through lack of action this usually results in reinforcing the pricing distortions referred to above.

Again, the young Ministry of Energy often has an inadequate budget and may lack the power to devote funds to timely energy projects or to determine energy taxes, subsidies, extended credit facilities and pricing policies or other instruments of financial management control over the energy mixture. Usually, the Ministry is so stretched professionally with day to day management imperatives that it also lacks the time to develop a comprehensive energy plan for the country as an essential starting point for implementing all the recommendations in this report. Furthermore, it is unable to develop the marketing of energy services, as opposed to fuels, in various sectors of the energy economy. Nor is it usually able to promote energy saving by fostering such important conservation procedures as energy audits, energy intensity analyses or thermal insulation of buildings.

All these shortcomings are so serious in terms of prudent energy management that governments should give the highest priority to establishing a top-level agency to plan and manage energy supply and demand of all types of energy in all sectors of national energy.

Recommendation:

We recommend that each nation designate a lead agency or authority with a broad mandate for the development of energy policy and its co-ordination, covering all energy sources and uses, and with clear responsibility to assess and to take into account the effects of energy strategies on economic growth and on sustaining the environmental basis of that growth. All single fuel agencies, including nuclear power, should be subordinate to its advice.

This lead agency should also be required to effect meaningful co-ordination with finance, economic development, environment, transport, agriculture, and other relevant ministries, and vice versa.

To be effective the agency must have access to some level of fiscal authority and become a primary source of advice on energy consumption taxes, subsidies, credit and other financial instruments used to influence energy development in favour of sustainable supply and consumption patterns.

This agency should have a special mandate to remove institutional barriers to the transformation of electrical supply and distribution utilities into *energy service* utilities, responsible for purchasing supplies from all sources, including renewables, at remunerative prices; and for marketing these supplies, tailored to specific end-uses in households, industry, agriculture, and other sectors.

In developing countries, it is essential that these agencies develop realistic national energy plans, and that any proposals or demand for foreign aid for energy development be made within the context of such a plan.

Regional Energy Commissions

Air, river or marine pollution from the waste products of energy production may cross national boundaries and cause problems for adjacent states of a world region, leading to damage and political embarrassment. The acid rain problem is a well known example of this, as are certain effluent accidents from nuclear installations. Another such issue is the regional identification of sites for the disposal of nuclear waste. They usually require collaboration by adjacent states for their solution which may involve mutually agreed or harmonized abatement and emission control codes of practice and standards.

Where many states are involved, detailed policy optimization practices based on overall least-cost solutions need to be worked out. Surveillance, monitoring and early warning practices developed by member states of the region could be made conformable and inter-comparable so that regional exchange of comparable data can be used to strengthen regional management.

On the more positive side, individual states with similar energy management problems may develop specialized technical skills for energy planning, hardware production or other expertise which they could share more easily if regional energy information exchange networks were developed. The activities of the SADCC (Southern African Development Coordination Conference) Regional Energy Centre in Luanda is a good example of a step in this direction.

In fact, the advantages of regional energy commissions are so numerous that a series of such commissions should be developed globally to carry out the above activities and similar collaborative functions.

Recommendation:

We recommend that states sharing common borders or similar socio-economic and bio-climatic conditions strengthen existing and/or establish new regional organizations with broad mandates for regional co-operation and joint action to deal with inter-related economic, energy, environment and development problems and to manage the energy transition in a co-ordinated and cost-effective way. These agencies would provide a much needed capability to identify and seize opportunities for regional co-operation in financing, developing and exploiting new technologies for energy supply, energy saving and environmental regeneration. They would also enable nations to develop regionally comparable economic and environmental statistics; baseline quantity and quality surveys of shared resources and an early warning capability to reduce and/or prevent an increasing range of environmental and developmental hazards.

Global Energy Environment Commission

Most energy-related problems, like the energy policies which induce them, reach across national boundaries, becoming regional and sometimes even global problems. Consequently, some form of international action is require for their resolution. Moreover, it is often more efficient for a number of countries with similar geographical, economic and other characteristics to join forces and tackle regional problems in a regional forum.

International action, however, cannot be and is not a substitute for national action. In general, effective international action should be based on similar and parallel action so that at national level.

It is abundantly clear from the discussion above, that no solution to these problems will be possible unless developing countries have access to assured finance from international sources. Now that the World Bank is concluding its energy accounting activities for developing countries, the idea of a *World Bank Energy Affiliate*, or some similar organization, with specific concerns for developing countries is more relevant than ever. Such a bank should increase overall energy-related lending over present levels, with particular emphasis on projects involving energy efficiency measures.

The line of argument developed in the previous sections could be extended to global problems such as the carbon dioxide and climate-warming issues or the release of radioactive gases to the atmosphere. These and similar global problems should form the basis for the creation of a Global Energy-Environment Commission whose task should be the development of internationally agreed abatement and/or adaptation strategies which would be developed in tandem with policy development. The work of such a Global Commission would be greatly facilitated by liaison with the Regional Energy Commissions mentioned above.

Recommendation:
At the global level also, we recommend that existing institutional capacity be significantly strengthened in order to enable nations to deal effectively with a growing range of energy-based environmental and developmental problems that transcend regional boundaries. These include: climatic change; the acidification of the environment; air pollution,; marine and coastal water pollution; and the problems posed by heavy metals, fuelwood, deforestation and so on. We recommend a three-track approach to these issues:

The first involves an acceleration of monitoring and assessment to improve our knowledge, strengthen our capacity for anticipation, prevention and adaptation and reduce the level of uncertainty. To this end existing arrangements for co-ordinated monitoring, assessment and research in the UN system need to be streamlined and broadened to include other sectors (e.g. agriculture, transportation, urban development) and other regions and countries likely to be severely hit by acidification or by changes in climate, rising sea levels, or other phenomena.

The second involves policies to promote research at different levels on energy and related strategies needed for the transition. These include: more effective instruments for influencing energy demand and overcoming barriers to their implementation; more energy-efficient means of heating, transportation and industrial development; efficiency standards for vehicles, equipment and appliances in international trade; alternative sources of energy; the economic costs of health, property and the environmental effects of alternative sources of energy; the geo-political implications of energy-induced climatic change, etc.

The third involves an immediate start on negotiations for an international convention on critical environmental and developmental problems related to energy, including those mentioned above. To this end, the Secretary General of the United Nations should convene a special conference not later than 1989, and charge it with drafting such a convention, which would be potentially as embracing and as binding as that of the Law of the Sea.

4. Annexes

Members of the Advisory Panel

Chairman:

Mr. Enrique IGLESIAS (Uruguay) — Minister for Foreign Affairs

Members:

Mr. Abdlatif Y. AL-HAMAD (Kuwait) —Director General, Chairman of the Board of Directors Arab Fund for Economic and Social Development

Mr. Toyoaki IKUTA (Japan) —President, Energy Economics Institute, Tokyo

Mr. GU Jian (China) —Chief Engineer, Wuhan Energy Research Institute

Mr. Al Noor KASSUM (Tanzania) —Minister for Water, Energy, and Minerals

Mr. Ulf LANTZKE (FRG)* — Former Director General, International Energy Agency, Paris

Mrs. Wangari MAATHAI (Kenya) —Chairman: National Council of Women, Green Belt Movement

Mr. David J. ROSE (USA)* — Professor, Massachusetts Institute of Technology

---

*Died during the period of the Panel's activities

Mr. Prem SHANKER JHA (India) — Senior Assistant Editor, The Times of India

Mr. Carl THAM (Sweden) — Director General, Swedish International Development Authority

Mr. György VAJDA (Hungary) — Director, Electrical Power Research Institute; Member, Hungarian Academy of Sciences

Mr. Miguel WIONCZEK (Mexico) — Director, Programme on Energy & Development

Special Adviser:

Professor Gordon T. Goodman, Director, The Beijer Institute, Stockholm, Sweden

Secretariat:

Kazu Kato, Programme Director
Janos Pasztor, Senior Programme Officer
Guadalupe Quesada, Secretary

Notes and References

1 FAO—1981
2 Keepin—1985
3 IIASA—1981
4 The group consisted of Jose Goldemberg of Brazil, Thomas B. Johansson of Sweden, Amulya K.N. Reddy of India, and Robert H. Williams of the USA. The initial findings of the group have been published in Goldemberg et al,—1985a
5 World Bank—1985
6 World Bank—1983
7 World Bank—1983
8 Goldemberg—1985b
9 The study was made by H. Geller of the Companhia Energetica de Sao Paulo, Brazil. Quoted by Goldemberg—1985b
10 Goldemberg—1985b
11 For more detailed discussion see the background papers Lohani—1985, Weidner—1985, Hashimoto—1985, CETESB—1985
12 For more detailed discussion see the background papers Torrens—1985, Zhao—1985, Rodhe—1985 and Goodman—1985
13 For more detailed discussion see the conference report Villach—1985, and the contributions to various public hearings of the WCED, such as Mintzer—1985 and Hare—1986
14 For a detailed discussion of the state of acidification in various European countries see WRI—1986
15 WRI—1986
16 Rodhe—1985
17 Quoted from page 224 of WRI—1986
18 Villach—1985
19 Goldemberg—1985b
20 Goldemberg—1985b

21 Deudney & Flavin—1983
22 There exists a large amount of literature on this. See for example: Chidumayo-1985, Ambio-1985, Beijer—1985-86.
23 See for example, Bandyopadhyay—1986
24 See for example Fernandes & Kulkarni—1983, Chidumayo—1985, Bradley—1985.
25 For a discussion of the potentials of nuclear energy see: IAEA—1985
26 For a more detailed discussion of this issue see Kasperson—1986.
27 Parker, et al.—1984, and IAEA—1983, etc.
28 Castaing—1983
29 For example, a number of countries supplying nuclear reactors also supply the nuclear fuel, which after use is returned to the supplier country for recycling or direct disposal. This happens with nuclear reactors and fuel supplied to the CMEA (Comecon) countries by the USSR.
30 For a more detailed analysis see Egan—1986.
31 MIT—1984

Bibliography

Special issue of **AMBIO, A Journal of the Human Environment**, devoted to *Energy in Developing Countries*, Royal Swedish Academy of Sciences/Pergamon Press, Volume XIV, Number 4-5, 1985.

BANDYOPADHYAY—1986 Jayanta Bandyopadhyay: *Rehabilitation of Upland Watersheds.* Paper commissioned by the WCED Secretariat, 1986.

BEIJER—1985-86 *Energy, Environment and Development in Africa*, Volumes 1-8. Published jointly by the Beijer Institute, Stockholm, and the Scandinavian Institute of African Studies, Uppsala.

BP-1985 *BP Statistical Review of World Energy,* British Petroleum, UK, June 1985.

BRADLEY—1985 Bradley et al.: *Development Research and Energy Planning in Kenya,* **AMBIO**, 1985, op cit.

CASTAING—1983 Report of the *Castaing Commission*. Rapport du groupe de travail sur les récherches et développement en matière de gestion des déchets radioactifs proposé par le CEA, 18 Mars, 1983, Paris.

CHIDUMAYO—1985 E.N. Chidumayo: Fuelwood and Social Forestry, P.O.Box: 50042, Lusaka, Zambia, Paper commissioned by the WCED Secretariat.

DEUDNEY & FLAVIN—1983 Daniel Deudney & Christopher Flavin: Renewable Energy, The power to choose. A Worldwatch Institute Book, Norton & Co., New York, 1983.

EGAN—1986 Joseph R. Egan: Life After Death for Nuclear Power, A global survey of new developments. Commissioned by the WCED Secretariat, January 1986.

FAO—1981 Map of the Fuelwood Situation in the Developing Countries, FAO, Rome, 1981.

FERNANDES & KULKARNI—1983 Walter Fernandes and Sharad Kulkarni, eds.: Towards a New Forest Policy: People's Rights and Environmental Needs. Indian Social Institute, New Delhi, 1983.

GOLDEMBERG et al.—1985a José Goldemberg et al.: An End-use Oriented Global Energy Strategy, Annual Review of Energy 1985, 10:613-88.

GOLDEMBERG et al.—1985b José Goldemberg et al.: Basic Needs and Much More with One Kilowatt per Capita, AMBIO—1985.

IAEA—1985 Energy and Nuclear Power Planning in Developing Countries, Technical Report Series No. 245, IAEA, Vienna.

IIASA—1981 Wolfgang Häfele et al.: Energy in a Finite World, Paths to a Sustainable Future, Ballinger, 1981.

KASPERSON—1986 Roger E. Kasperson: The Public Acceptance of Nuclear Energy: Looking to the year 2000 and beyond. Paper commissioned by the WCED Secretariat, March 1986.

KEEPIN—1985 Bill Keepin et al.: Future Energy and CO2 Projections, The Beijer Institute, Stockholm, Sweden, 1985.

MILLER—1986 Miller, Mintzer and Hoagland: Growing Power: Bioenergy for development and industry. World Resources Institute, Study 5, April 1986.

MINTZER—1985 Statement by Dr, Irving Mintzer of the World Resources Institute to the Oslo Public Hearing of the World Commission on Environment and Development, June 24, 1985.

MIT—1984 The Future of the Automobile. The Report of MIT's International Automobile Program. George Allen & Unwin, London, 1984.

OECD—1985 The State of the Environment 1985, OECD, Paris 1985.

PARKER et al.—1984 Parker, Broshears and Pasztor: The Disposal of High-level Radioactive Waste—1984., Vols. I and II, The Beijer Institute, Stockholm, Sweden.

RODHE—1985 Henning Rodhe: Acidification in Tropical Countries. Paper Commissioned by the WCED Secretariat, June 1985.

TORRENS—1985 Ian M. Torrens: Acid Rain and Air Pollution: A problem of industrialization. Paper commissioned by the WCED Secretariat, June 1985.

VILLACH—1985 The Report of the International Conference organized jointly by ICSU, WMO and UNEP.

WORLD BANK—1983 The Energy Transition in Developing Countries, The World Bank, Washington DC, USA, 1983.

WRI—1986 World Resources—1986. An assessment of the Resource Base that Supports the Global Economy. World Resources Institute, Washington DC., May 1986.

Glossary

Acidific: general term describing the phenomena by which various oxides (principally of sulphur and nitrogen) resulting from the combustion of fossil fuels, are transformed into acids in the atmosphere and deposited in a wet or dry form on the ground, in lakes or on plants at a certain distance from the source.

Agroforestry: a collective name for land-use systems and technologies where woody perennials (trees, shrubs etc.) are deliberately used on the same land management unit as agricultural crops and/or animals, either in some form of spatial arrangement or temporal sequence. In agroforestry systems there are both ecological and economic interactions between different components.

Biomass: general term for all living matter. Biomass energy sources include living matter which can be directly used or converted into fuel.

Biomass Wastes: general term for living matter destined for waste. Examples include wood chips, animal dung, crop wastes, etc. Most biomass wastes can be converted into useful fuels.

Cogeneration: the process by which waste heat produced in a thermal or nuclear power plant is partially recovered and used for central heating, etc. In such systems the total amount of electricity produced is somewhat less than in simple systems, but given the additional useful heat produced, the overall thermal efficiency is much higher.

Conversion: the process by which one form of energy is converted into another, usually a more useful form. Thus the heat energy liberated from the combustion of coal is converted to electricity in a thermal power plant. Similarly, the energy stored in biomass is converted to biogas in a biogas plant.

Decommissioning: the process by which a nuclear reactor, the life of which is limited to 25-45 years, is disassembled, and the radioactive parts decontaminated or disposed of.

Dedicated Energy Plantation: a biomass plantation, usually trees, whose purpose is to supply wood for energy, often for a specific industry or city.

End-Use: the point at which, after a number of conversions from one fuel to another, energy is used up, i.e., by the user to do useful work.

Eutrophication: the process of nutrient enrichment of water which leads to increased organic growth but which, if carried too far (hypertrophication), causes undesirable results.

Exajoule: 10^{18} joules (one billion billion joules).

Fuelwood: wood used for direct combustion.

Gasification: the conversion of a range of biomass (including wood, charcoal, crop residues, etc.) into a combustible gas under oxygen-free conditions.

Geological Disposal: the process of storing radioactive wastes in geological formations, usually in non-retrievable configurations.

Gigawatts: one billion watts, or GW, or 10^9 watts (see Joules).

High-level: applied to radioactive waste, this applies to substances which are so radioactive that they generate heat. Other groups include medium—or intermediate—and low-level. The classification is approximate, and is only for convenience.

Joule: the standard measure of energy, defined as the work done by a force of one newton when it moves its point of application over a distance of one metre. A system doing work at the rate of one joule per second has a power of one *watt*. As the joule is of small magnitude, it is generally used as a multiple, such as gigajoule (GJ, or 10^9 joules, or a billion joules), terajoules (TJ, 10^{12} joules, or a trillion joules), or exajoules (EJ, or 10^{18} joules, or a million trillion joules).

Kilowatts: one thousand watts (see joules), or KW, or 10^3 watts.

Long-Lived: radionuclides with long half-lives. The division is arbitrary, but generally includes radionuclides with half-lives of hundreds of years and over.

Megawatts: one million watts (see joule), or MW, or 10^6 watts.

Non-Renewable: used of an energy source where the rate of use is higher than the sustainable rate of production, e.g., fossil fuels which took millions of years in their formation and are being used up in a matter of a few hundred years.

Nuclear Fuel Cycle: includes all stages of mining uranium ore, enrichment, fuel manufacture, burn-up in a reactor, reprocessing or conditioning, interim storage, and final disposal.

Proven Reserves: generally used of fossil-fuel sources to indicate those reserves of oil, coal or gas which have been found. It does not include those reserves which are likely to exist.

75

Terawatts: trillion watts (see joules also), or TW, or 10^{12} watts.

Titration: the process of determining the strength of a solution or the concentration of a substance in solution in terms of the smallest amount of reagent of known concentration required to bring about a given effect in reaction with a known volume of the test solution.

Transmutation: the conversion of one radioactive nuclide into another by neutron irradiation. The purpose of such transmutation is to change nuclides with very long half-lives into others with much shorter ones.

Waste
Management: all aspects of managing the concluding stages of the nuclear fuel cycle, i.e. fuel conditioning, reprocessing, interim storage, and final disposal.

Wood Fuels: all fuels made from wood. The simplest forms are fuelwood and charcoal. Other, more advanced fuels such as producer gas, methanol, and even ethanol can also be made from wood.